QUANTUM-MECHANICAL PREDICTION OF THERMOCHEMICAL DATA

Understanding Chemical Reactivity

Volume 22

The titles published in this series are listed at the end of this volume.

Quantum-Mechanical Prediction of Thermochemical Data

edited by

Jerzy Cioslowski
Florida State University,
Tallahassee, Florida, U.S.A.

KLUWER ACADEMIC PUBLISHERS
DORDRECHT / BOSTON / LONDON

A C.I.P. Catalogue record for this book is available from the Library of Congress.

ISBN 1-4020-0424-9

Published by Kluwer Academic Publishers,
P.O. Box 17, 3300 AA Dordrecht, The Netherlands.

Sold and distributed in North, Central and South America
by Kluwer Academic Publishers,
101 Philip Drive, Norwell, MA 02061, U.S.A.

In all other countries, sold and distributed
by Kluwer Academic Publishers,
P.O. Box 322, 3300 AH Dordrecht, The Netherlands.

Printed on acid-free paper

Table of Contents

Preface xi

Contributors xiii

Chapter 1. Highly Accurate Ab Initio Computation of Thermochemical Data

Trygve Helgaker, Wim Klopper, Asger Halkier, Keld L. Bak, Poul Jørgensen and Jeppe Olsen

1. Introduction 1
2. Hierarchies of Ab Initio Theory 2
   2.1. The Coupled-Cluster Hierarchy of n-Electron Models 2
   2.2. The Correlation-Consistent Hierarchy of One-Electron Basis Sets 4
   2.3. Computational Cost 5
3. Convergence of the Coupled-Cluster Hierarchy 5
   3.1. Model Calculations on N_2 and HF 5
   3.2. The CCSD(T) Model 7
4. An Illustrative Example: the Atomization Energy of CO 8
   4.1. Electronic and Nuclear Contributions 9
   4.2. Dependence on the AO Basis Set 11
5. Short-Range Correlation and the Coulomb Hole 12
   5.1. Terms Linear in r_{12} 12
   5.2. Extrapolations from Principal Expansions 15
6. Calibration of the Extrapolation Technique 16
   6.1. Valence-Shell Correlation Energy 16
   6.2. Total Electronic Energy 19
   6.3. Core Contributions to AEs 22
7. Molecular Vibrational Corrections 22
8. Relativistic Contributions 24
9. Calculation of Atomization Energies 25
10. Conclusions and Perspectives 25

References 28

Chapter 2. W1 and W2 Theories, and Their Variants: Thermochemistry in the kJ/mol Accuracy Range

Jan M.L. Martin and S. Parthiban

1. Introduction and Background 31
2. Steps in the W1 and W2 Theories, and Their Justification 33
   2.1. Reference Geometry 34
   2.2. The SCF Component of TAE 35
   2.3. The CCSD Valence Correlation Component of TAE 38
   2.4. Connected Triple Excitations: the (T) Valence Correlation Component of TAE 39
   2.5. The Inner-Shell Correlation Component of TAE 40
   2.6. Scalar Relativistic Correction 41
   2.7. Spin-Orbit Coupling 42
   2.8. The Zero-Point Vibrational Energy 43
3. Performance of W1 and W2 theories 46
   3.1. Atomization Energies (the W2-1 Set) 46
   3.2. Electron Affinities (the G2/97 Set) 48
   3.3. Ionization Potentials (the G2/97 Set) 48
   3.4. Heats of Formation (the G2/97 Set) 50
   3.5. Proton Affinities 50
4. Variants and Simplifications 50
   4.1. W1′ Theory 50
   4.2. W1h and W2h Theories 51
   4.3. A Bond-Equivalent Model for Inner-Shell Correlation 52
   4.4. Reduced-Cost Approaches to the Scalar Relativistic Correction 54
   4.5. W1c Theory 56
   4.6. Detecting Problems 56
5. Example Applications 57
   5.1. Heats of Vaporization of Boron and Silicon 57
   5.2. Validating DFT Methods for Transition States: the Walden Inversion 58
   5.3. Benzene as a "Stress Test" of the Method 59
6. Conclusions and Prospects 61
References 62

Chapter 3. Quantum-Chemical Methods for Accurate Theoretical Thermochemistry

Krishnan Raghavachari and Larry A. Curtiss

1. Introduction 67

2. The G3/99 Test Set 69
3. Gaussian-3 Theory 70
4. G3S Theory 77
5. G3X Theory 81
6. Density Functional Theory 88
7. Concluding Remarks 94
References 95

Chapter 4. Complete Basis Set Models for Chemical Reactivity: from the Helium Atom to Enzyme Kinetics

George A. Petersson

1. Introduction 99
2. Pair Natural Orbital Extrapolations 100
3. Current CBS Models 102
4. Transition States 104
5. Explicit Functions of the Interelectron Distance 109
6. The cc-pVnZ Basis Sets 110
7. New Developments 112
   7.1. The SCF Limit 113
   7.2. The CBS Limit for the MP2 Correlation Energy 114
   7.3. The Higher-Order Correlation Energy 117
   7.4. Total Energies 118
8. Enzyme Kinetics and Mechanism 120
9. Summary 127
References 127

Chapter 5. Application and Testing of Diagonal, Partial Third-Order Electron Propagator Approximations

Antonio M. Ferreira, Gustavo Seabra, O. Dolgounitcheva, V. G. Zakrzewski, and J. V. Ortiz

1. Introduction 131
2. Electron Propagator Concepts 132
3. An Economical Approximation: P3 134
4. Other Diagonal Approximations 138
5. Nondiagonal Approximations 140
6. An Example of Application of P3: 9-Methylguanine 141
7. P3 Test Results 145
   7.1. Atomic Ionization Energies 145
   7.2. Molecular Species 151

8. Conclusions and Prospectus 155
References 156

Chapter 6. Theoretical Thermochemistry of Radicals

David J. Henry and Leo Radom

1. Introduction 161
2. Theoretical Procedures 162
3. Geometries 167
4. Heats of Formation 169
5. Bond Dissociation Energies 174
6. Radical Stabilization Energies 177
7. Reaction Barriers 181
8. Reaction Enthalpies 191
9. Concluding Remarks 193
References 194

Chapter 7. Theoretical Prediction of Bond Dissociation Energies for Transition Metal Compounds and Main Group Complexes with Standard Quantum-Chemical Methods

Nikolaus Fröhlich and Gernot Frenking

1. Introduction 199
2. Homoleptic Carbonyl Complexes 203
3. Group-6 Carbonyl Complexes $M(CO)_5L$ (M = Cr, Mo, W) 206
4. Iron Carbonyl Complexes $Fe(CO)_4L$ 207
5. Group-10 Carbonyl Complexes $M(CO)_3L$ (M = Ni, Pd, Pt) 209
6. Group-6 Carbonyl Complexes with Phosphane Ligands $M(CO)_5PR_3$ (M = Cr, Mo, W; R = H, Me, F, Cl) 210
7. Noble Gas Complexes $M(CO)_5Ng$ (M = Cr, Mo, W; Ng = Ar, Kr, Xe) 210
8. Transition Metal Carbene and Carbyne Complexes 211
9. Transition Metal Complexes with π-Bonded Ligands 214
10. Transition Metal Complexes with Group-13 Diyl Ligands ER (E = B, Al, Ga, In, Tl) 216
11. Transition Metal Compounds with Boryl Ligands BR_2 and Gallyl Ligands GaR_2 220
12. Transition Metal Methyl and Phenyl Compounds 221
13. Transition Metal Nitrido and Phosphido Complexes 222
14. Main Group Complexes of Group-13 Lewis Acids EX_3 (E = B - Tl; X = H, F, Cl) 224
15. Main Group Complexes of BeO 226

16. Conclusions 228
References 229

Chapter 8. Theoretical Thermochemistry: a Brief Survey

Walter Thiel

1. Introduction 235
2. Theoretical Background 236
3. Specific Conventions 237
4. Statistical Evaluations 238
5. Discussion 242
References 244

Index 247

Preface

For the first time in the history of chemical sciences, theoretical predictions have achieved the level of reliability that allows them to rival experimental measurements in accuracy on a routine basis. Only a decade ago, such a statement would be valid only with severe qualifications as high-level quantum-chemical calculations were feasible only for molecules composed of a few atoms. Improvements in both hardware performance and the level of sophistication of electronic structure methods have contributed equally to this impressive progress that has taken place only recently.

The contemporary chemist interested in predicting thermochemical properties such as the standard enthalpy of formation has at his disposal a wide selection of theoretical approaches, differing in the range of applicability, computational cost, and the expected accuracy. Ranging from high-level treatments of electron correlation used in conjunction with extrapolative schemes to semiempirical methods, these approaches have well-known advantages and shortcomings that determine their usefulness in studies of particular types of chemical species. The growing number of published computational schemes and their variants, testing sets, and performance statistics often makes it difficult for a scientist not well versed in the language of quantum theory to identify the method most adequate for his research needs.

In this book, the experts who have developed and tested many of the currently used electronic structure procedures present an authoritative overview of the theoretical tools for the computation of thermochemical properties of atoms and molecules. The first two chapters describe the highly accurate, computationally expensive approaches that combine high-level calculations with sophisticated extrapolation schemes. In chapters 3 and 4, the widely used G3 and CBS families of composite methods are discussed. The applications of the electron propagator theory to the estimation of energy changes that accompany electron detachment and attachment processes follow in chapter 5. The next two sections of the book focus on practical applications of the aforedescribed methods to free radicals and organometallic compounds. Finally, a brief review of semiempirical methods is given in chapter 8.

Since the science presented here would never materialize without productive interactions between theory and experiment, it is certainly appropriate to dedicate this book to the practitioners of experimental chemistry who do not hesitate to regard electronic structure calculations as an integral part of their investigations and to the vanguards of molecular quantum mechanics who do not shy away from visiting research laboratories where matter rather than its abstract representations is studied.

Jerzy Cioslowski

Tallahassee, April 2001

Contributors

Keld L. Bak, UNI-C, Olof Palmes Allé 38, DK-8200 Århus N, Denmark

Larry A. Curtiss, Materials Science and Chemistry Divisions, Argonne National Laboratory, Argonne, IL 60439, U.S.A.

O. Dolgounitcheva, Department of Chemistry, Kansas State University, Manhattan, Kansas 66506-3701

Antonio M. Ferreira, Department of Chemistry, Kansas State University, Manhattan, Kansas 66506-3701

Gernot Frenking, Fachbereich Chemie, Philipps-Universität Marburg, Hans-Meerwein Strasse, D-35032, Marburg, Germany

Nikolaus Fröhlich, Fachbereich Chemie, Philipps-Universität Marburg, Hans-Meerwein Strasse, D-35032, Marburg, Germany

Asger Halkier, Theoretical Chemistry Group, Debye Institute, Utrecht University, P. O. Box 80052, NL-3508 TB Utrecht, The Netherlands

Trygve Helgaker, Department of Chemistry, University of Oslo, P.O. Box 1033 Blindern, N-0315 Oslo, Norway

David J. Henry, Research School of Chemistry, Australian National University, Canberra, ACT 0200, Australia

Poul Jørgensen, Department of Chemistry, Århus University, DK-8000 Århus C, Denmark

Wim Klopper, Theoretical Chemistry Group, Debye Institute, Utrecht University, P. O. Box 80052, NL-3508 TB Utrecht, The Netherlands

Jan M.L. Martin, Department of Organic Chemistry, Weizmann Institute of Science, Kimmelman Building, IL-76100 Reḥovot, Israel

Jeppe Olsen, Department of Chemistry, Århus University, DK-8000 Århus C, Denmark

J. V. Ortiz, Department of Chemistry, Kansas State University, Manhattan, Kansas 66506-3701

S. Parthiban, Department of Organic Chemistry, Weizmann Institute of Science, Kimmelman Building, IL-76100 Reḥovot, Israel

George A. Petersson, Hall-Atwater Laboratories of Chemistry, Wesleyan University, Middletown, Connecticut 06459-0180, U.S.A.

Leo Radom, Research School of Chemistry, Australian National University, Canberra, ACT 0200, Australia

Krishnan Raghavachari, Electronic and Photonic Materials Physics Research, Agere Systems, Murray Hill, NJ 07974, U.S.A.

Gustavo Seabra, Department of Chemistry, Kansas State University, Manhattan, Kansas 66506-3701

Walter Thiel, Max-Planck-Institut für Kohlenforschung, Kaiser-Wilhelm-Platz 1, D-454 70 Mülheim, Germany

V. G. Zakrzewski, Department of Chemistry, Kansas State University, Manhattan, Kansas 66506-3701

Chapter 1

Highly Accurate Ab Initio Computation of Thermochemical Data

Trygve Helgaker
Department of Chemistry, University of Oslo, P. O. Box 1033 Blindern, N-0315 Oslo, Norway

Wim Klopper and Asger Halkier
Theoretical Chemistry Group, Debye Institute, Utrecht University, P. O. Box 80052, NL-3508 TB Utrecht, The Netherlands

Keld L. Bak
UNI-C, Olof Palmes Allé 38, DK-8200 Århus N, Denmark

Poul Jørgensen and Jeppe Olsen
Department of Chemistry, Århus University, DK-8000 Århus C, Denmark

1. INTRODUCTION

Heats of reaction are among the fundamental quantities of thermochemistry. Since, to a first approximation, the heats of reaction are energy differences between molecular systems, one would think that their quantum-chemical evaluation should be a rather straightforward matter. After all, ever since its inception in the late 1920s, quantum chemistry has been concerned with the accurate calculation of total energies, building up a large body of expertise and experience on the accurate and efficient calculation of total molecular electronic energies [1-10].

J. Cioslowski (ed.), Quantum-Mechanical Prediction of Thermochemical Data, 1–30.

Nevertheless, in spite of all these efforts, it has proved exceedingly difficult to compute the necessary quantities (i.e., total electronic energies) to a target accuracy of better than 1 kJ/mol, which is the typical accuracy of experimental measurements of heats of formation. In this chapter, we examine the ab initio calculation of atomization energies (AEs) of gas-phase molecules, from which the heats of gas-phase reactions between the same molecules can be easily obtained. Our purpose is not only to illustrate the inherent difficulties associated with the accurate calculation of AEs, but also to describe the considerable progress that has been achieved over the last few years and the perspectives for the near future.

2. HIERARCHIES OF AB INITIO THEORY

An important characteristic of ab initio computational methodology is the ability to approach the exact description – that is, the *focal point* [11] – of the molecular electronic structure in a systematic manner. In the standard approach, approximate wavefunctions are constructed as linear combinations of antisymmetrized products (determinants) of one-electron functions, the molecular orbitals (MOs). The quality of the description then depends on the basis of atomic orbitals (AOs) in terms of which the MOs are expanded (the one-electron space), and on how linear combinations of determinants of these MOs are formed (the n-electron space). Within the one- and n-electron spaces, hierarchies exist of increasing flexibility and accuracy. To understand the requirements for accurate calculations of thermochemical data, we shall in this section consider the one- and n-electron hierarchies in some detail [12].

2.1. The Coupled-Cluster Hierarchy of n-Electron Models

At the lowest level of the n-electron hierarchy, we have the Hartree-Fock wavefunction, obtained by variationally optimizing a single determinant with respect to the shape of the occupied MOs. As we shall see, the quality of the single-determinant Hartree-Fock description is too low to provide sufficiently accurate thermochemical data. The inadequacy of the Hartree-Fock model arises from the fact that a single determinant gives an uncorrelated description of the electronic motion, in which each electron moves in the average (rather than instantaneous) field generated by the other electrons. This model is therefore incapable of describing the subtle changes that occur as electron pairs are broken or formed during chemical processes. For an accurate description of thermochemical data, we must therefore go beyond the Hartree-Fock level of theory.

A standard method of improving on the Hartree-Fock description is the coupled-cluster approach [12, 13]. In this approach, the wavefunction $|\mathrm{CC}\rangle$ is written as an exponential of a cluster operator $\hat{\mathrm{T}}$ working on the Hartree-Fock state $|\mathrm{HF}\rangle$, generating a linear combination of all possible determinants that may be constructed in a given one-electron basis,

$$|\mathrm{CC}\rangle = \exp(\hat{\mathrm{T}})\,|\mathrm{HF}\rangle. \tag{2.1}$$

The cluster operator $\hat{\mathrm{T}}$ creates excitations out of the Hartree-Fock determinant and may be written as

$$\hat{\mathrm{T}} = \hat{\mathrm{T}}_1 + \hat{\mathrm{T}}_2 + \hat{\mathrm{T}}_3 + \cdots, \tag{2.2}$$

where $\hat{\mathrm{T}}_1$ creates single excitations, $\hat{\mathrm{T}}_2$ creates double excitations, and so on. There is ample theoretical and numerical evidence that the contributions from $\hat{\mathrm{T}}_\mathrm{i}$ rapidly decrease after $\mathrm{i} = 2$ (double excitations), and the coupled-cluster operator is therefore usually truncated either after $\mathrm{i} = 2$, leading to the coupled-cluster singles-and-doubles (CCSD) model, or after $\mathrm{i} = 3$, leading to the CCSDT method, including all singles, doubles, and triples. The advantage of the exponential parameterization is that the wavefunction becomes multiplicatively separable, thereby providing a uniform (size-extensive) description of systems of different size. Other hierarchies of n-electron models exist, but the coupled-cluster hierarchy has proved to be the most successful for highly accurate calculations of molecular electronic structure. In particular, the configuration-interaction (CI) method suffers from a lack of size-extensivity and the related slow convergence with respect to the level of excitation; the perturbative Møller-Plesset method is not sufficiently accurate to low orders and converges slowly if at all [14].

To understand the structure of the coupled-cluster wavefunction, let us Taylor expand the exponential in Eq. (2.1). Sorting the resulting expansion according to the level of excitation, we obtain

$$\begin{aligned} \exp(\hat{\mathrm{T}})\,|\mathrm{HF}\rangle &= |\mathrm{HF}\rangle + \hat{\mathrm{T}}_1\,|\mathrm{HF}\rangle \\ &+ \left(\hat{\mathrm{T}}_2 + \tfrac{1}{2}\hat{\mathrm{T}}_1^2\right)|\mathrm{HF}\rangle \\ &+ \left(\hat{\mathrm{T}}_3 + \hat{\mathrm{T}}_1\hat{\mathrm{T}}_2 + \tfrac{1}{6}\hat{\mathrm{T}}_1^3\right)|\mathrm{HF}\rangle + \cdots . \end{aligned} \tag{2.3}$$

At each excitation level beyond the single-excitation level, a number of terms contribute. For example, double excitations are generated both by means of the double-excitation operator $\hat{\mathrm{T}}_2$ (connected excitations) and by means of two independent single-excitation operators $\hat{T}_1^2$ (disconnected excitations). At the CCSD level, we ignore all triple and higher connected excitations. Nevertheless, because of the presence of disconnected contributions in Eq. (2.3), the resulting CCSD wavefunction contains contributions from all levels of excitation – that is, it contains contributions from all determinants that can be constructed in a given AO basis. Since the higher excitations are dominated by disconnected contributions, this separation into connected and disconnected contributions ensures both a rapid convergence of the coupled-cluster hierarchy and size-extensivity of its energy.

2.2. The Correlation-Consistent Hierarchy of One-Electron Basis Sets

The quality of quantum-chemical calculations depends not only on the chosen n-electron model but also critically on the flexibility of the one-electron basis set in terms of which the MOs are expanded. Obviously, it is possible to choose basis sets in many different ways. For highly accurate, systematic studies of molecular systems, it becomes important to have a well-defined procedure for generating a sequence of basis sets of increasing flexibility. A popular hierarchy of basis functions are the correlation-consistent basis sets of Dunning and coworkers [15-17]. We shall use two varieties of these sets: the cc-pVXZ (correlation-consistent polarized-valence X-tuple-zeta) and cc-pCVXZ (correlation-consistent polarized core-valence X-tuple-zeta) basis sets; see Table 1.1.

As can be seen from the table, the number of AOs increases rapidly with the cardinal number X. Thus, with each increment in the cardinal number, a new shell of valence AOs is added to the cc-pVXZ set; since the number of AOs added in each step is proportional to X^2, the total number (N_{bas}) of AOs in a correlation-consistent basis set is proportional to X^3. The core-valence sets cc-pCVXZ contain additional AOs for the correlation of the core electrons. As we shall see later, the hierarchy of correlation-consistent basis sets provides a very systematic description of molecular electronic systems, enabling us to develop a useful extrapolation technique for molecular energies.

Table 1.1 Correlation-consistent basis sets for first-row atoms.

X	cc-pVXZ	N_{bas}	cc-pCVXZ	N_{bas}
D	3s2p1d	14	4s3p1d	18
T	4s3p2d1f	30	6s5p3d1f	43
Q	5s4p3d2f1g	55	8s7p5d3f1g	84
5	6s5p4d3f2g1h	91	10s9p7d5f3g1h	145
6	7s6p5d4f3g2h1i	140	12s11p9d7f5g3h1i	230

2.3. Computational Cost

The computational complexity of the coupled-cluster method truncated after a given excitation level m – for example, m = 2 for CCSD – may be discussed in terms of the number of amplitudes (N_{am}) in the coupled-cluster operator and the number of operations (N_{op}) required for optimization of the wavefunction. Considering K atoms, each with N_{bas} basis functions, we have the following scaling relations:

$$N_{am} \propto K^{2m} N_{bas}^{m} \propto K^{2m} X^{3m}, \tag{2.4}$$

$$N_{op} \propto K^{2m+2} N_{bas}^{m+2} \propto K^{2m+2} X^{3m+6}. \tag{2.5}$$

From these expressions, it is evident that the steep increase of the computational cost with X (the basis-set hierarchy) and m (the coupled-cluster hierarchy) severely restricts the levels of theory that can be routinely used for large systems or even explored for small systems. Nevertheless, we shall see that, with current computers, it is possible to arrange the calculations in such a manner that chemical accuracy (of the order of 1 kcal/mol $\approx$ 4 kJ/mol) can be achieved – at least for molecules containing not more than ten first-row atoms.

3. CONVERGENCE OF THE COUPLED-CLUSTER HIERARCHY

3.1. Model Calculations on N_2 and HF

For small basis sets and molecules, it is possible to calculate the full set of energies in the coupled-cluster hierarchy, from the Hartree-Fock to the full configuration-interaction (FCI) energy [18]. Although such

Table 1.2 Convergence of the coupled-cluster hierarchy for the valence-shell electronic energies (E_e) and equilibrium AEs of N_2 and HF in the cc-pVDZ basis. The energies are given in kJ/mol and are relative to the FCI values.

	N_2 [a]		HF [b]	
	ΔE_e [c]	ΔAE [d]	ΔE_e [e]	ΔAE [f]
HF	845.34	-372.03	549.32	-140.08
CCSD	35.35	-31.14	6.34	-3.45
CCSDT	4.27	-4.05	1.06	-0.74
CCSDTQ	0.50	-0.50	0.03	-0.01
CCSDTQ5	0.04	-0.04	0.00	0.00
CCSDTQ56	0.00	0.00		

[a] $R_e = 2.068$ a_0.
[b] $R_e = 0.91694$ Å.
[c] Relative to the valence FCI energy of -109.2765 E_h.
[d] Relative to the valence FCI AE of 838.65 kJ/mol.
[e] Relative to the valence FCI energy of -100.2286 E_h.
[f] Relative to the valence FCI AE of 529.37 kJ/mol.

calculations employ basis sets that are too small to give quantitative estimates of AEs, they still provide useful information about the convergence of the coupled-cluster hierarchy – in particular, about the importance of the quadruple and higher excitations. We here present two such series of calculations.

In Table 1.2, we have listed the valence cc-pVDZ electronic energies and AEs of N_2 and HF at different levels of coupled-cluster theory. The energies are given as deviations from the FCI values. Comparing the different levels of theory, we note that the error is reduced by one order of magnitude at each level. In particular, at the CCSDT level, there is a residual error of the order of a few kJ/mol in the calculated energies and AEs, suggesting that the CCSDTQ model is usually needed to reproduce experimental measurements to within the quoted errors bars (often less than 1 kJ/mol).

Let us make two more observations about the convergence of the coupled-cluster hierarchy. First, it converges faster for HF than for N_2, reflecting the more complicated electronic structure of the multiple bond in N_2. Second, the cancellation of errors in the calculated AEs becomes less pronounced as we move up in the coupled-cluster hierarchy. The cancellation diminishes since the coupled-cluster expansions of the atoms converge faster than those of the molecules, which contain more electron pairs.

Although the calculations reported here have been carried out in a small basis, there is no reason to believe that our conclusions regarding the convergence of the coupled-cluster hierarchy would be different had the calculations been carried out in larger basis. In particular, we conclude that the CCSDT model is incapable of predicting AEs to within 1 kJ/mol.

3.2. The CCSD(T) Model

As shown in the previous section, the coupled-cluster hierarchy converges rapidly, the error in the total error being reduced by an order of magnitude at each new level of theory. Unfortunately, from section 2.3, we recall that the cost of the coupled-cluster calculations increases very rapidly with the inclusion of higher-order connected excitations. In practice, while it is possible to carry out CCSD calculations for fairly large systems and basis sets (more than 10 atoms at the cc-pCVQZ level), the full CCSDT model is presently too expensive for routine calculations. However, since we are anyway forced to neglect the connected quadruples (CCSDTQ) in our calculations, the overall quality of our calculations will not be adversely affected if we make an approximation in the treatment of the connected triples whose error is not larger than that incurred by neglecting the quadruples. In practice, therefore, any approximate treatment of the triples that gives an error of the order of 10 % or less would be welcome.

Among the various approximate methods for including the connected triple excitations, the CCSD(T) method is the most popular [19]. In this approach, the CCSD calculation is followed by the calculation of a perturbational estimate of the triple excitations. In addition to reducing the overall scaling with respect to the number of atoms K from K^8 in CCSDT [see Eq. (2.5)] to K^7 in CCSD(T), the CCSD(T) method avoids completely the storage of the triples amplitudes.

In Table 1.3, we have listed the contributions from the single and connected double and triple excitations to the AEs of CH_2, CO, F_2, H_2O, HF, and N_2 at the valence-electron CCSDT/cc-pV5Z and CCSD(T)/cc-pV5Z levels [20]. The second column of Table 1.3 contains the CCSD singles and doubles contributions to the correlation energy, the third column the triples contributions as obtained in the CCSD(T) method, and the last column the difference between the triples contributions in CCSDT and CCSD(T) – that is, the energy contribution that originates from the full relaxation of the triples. The error incurred by employing

Table 1.3 Valence-shell contributions of connected double and triple excitations to the AEs of six molecules in the cc-pV5Z basis (kJ/mol).

	HF → SD	SD → (T)	(T) → T
CH_2 [a]	214.09	8.03	0.81
CO	312.65	33.29	-2.24
F_2	281.18	31.27	-1.28
H_2O	302.20	14.53	-0.91
HF	175.23	8.83	-0.62
N_2	419.22	38.96	-3.02

[a] $^1A_1$ state.

CCSD(T) instead of CCSDT amounts to no more than 10 % of the total triples correction and 1 % of the total correlation energy, thus fulfilling our requirement for an acceptable approximate triples theory.

However, the success of the CCSD(T) model stems not only from the fact that it gives a good approximation to the full triples correction. From Table 1.3, we note that the CCSD(T) model usually *overestimates* the contributions from the triples, the only exception being CH_2. This overestimation is particularly significant for N_2, where the CCSD(T) triples correction is 3 kJ/mol larger than the full triples correction. For comparison, it is seen from Table 1.2 that, in the cc-pVDZ basis, the connected quadruple excitations add 3.55 kJ/mol to the AE. The overestimation of the triples contribution by the CCSD(T) model will thus partly cancel the error incurred by ignoring the connected quadruples. In general, therefore, we may expect that the CCSD(T) AEs will not be improved by going to the full CCSDT method [20]. In this sense, the CCSD(T) model represents a very accurate method for the calculation of AEs, which may only be improved upon by simultaneously including more terms from the connected triples as well as contributions from the connected quadruples [20-22].

4. AN ILLUSTRATIVE EXAMPLE: THE ATOMIZATION ENERGY OF CO

To illustrate the difficulties associated with the accurate calculation of thermochemical data, we here consider the calculation of the AE of CO – that is, the difference in total energy between the CO molecule

Table 1.4 The ab initio calculation of the AE of the CO molecule (kJ/mol) at the basis-set limit. All calculations have been carried out at the all-electron CCSD(T)/cc-pCVQZ geometry [25].

	E_C	+	E_O	−	E_{CO}	=	AE
HF[a]	-98964.9	−	196437.1	+	296132.2	=	730.1
SD[b]	-388.4	−	639.6	+	1350.2	=	322.1
(T)[c]	-7.7	−	11.9	+	54.2	=	34.6
ZPVE[d]	0.0	+	0.0	−	12.9	=	-12.9
Rel.[e]	40.1	−	139.0	+	177.1	=	-2.0
Total	-99401.1	−	197227.6	+	297700.8	=	1072.0
Exp.[f]							1071.8

[a] RHF/cc-pV6Z for CO and UHF/cc-pV6Z for C and O.
[b] Calculated at the all-electron CCSD/cc-pcV(56)Z level.
[c] Calculated at the all-electron CCSD(T)/cc-pcV(56)Z level.
[d] Experimental value from Ref. 23.
[e] First-order relativistic (scalar and spin-orbit) corrections calculated at the all-electron CCSD(T)/cc-pCVQZ level.
[f] Experimental value from Ref. 24.

and the C and O atoms at 0 K. The experimental AE of CO is known to be 1071.8(5) kJ/mol [23, 24]; in Table 1.4, we have collected the various contributions to the theoretical AE of CO, as calculated at the CCSD(T) level in the limit of a complete one-electron basis. Note that the calculated AE of 1072.0 kJ/mol is within the experimental error bars, even though it constitutes less than 0.5 % of the total energy of the CO molecule.

4.1. Electronic and Nuclear Contributions

Let us discuss the various contributions to the calculated AE of CO. The first row of Table 1.4 contains the energies obtained from separate Hartree-Fock calculations on CO and its constituents. The large error of 32 % arises since the Hartree-Fock model is incapable of describing the complicated changes in the electronic structure that occur as electron pairs are broken. Although we might hope that the errors associated with the breaking of electron pairs to some extent cancel in the enthalpies of isogyric reactions (i.e., reactions that conserve the number of electron pairs), we clearly need to go beyond the one-determinant Hartree-Fock description for a satisfactory theoretical prediction of AEs.

In the second row of Table 1.4, we have listed the corrections to the Hartree-Fock energies that are obtained from CCSD calculations. Clearly, we now have a better description of the atomization process, the error in the calculated AE being only -19.6 kJ/mol (2 %). Still, we are far away from the prescribed target accuracy of 1 kJ/mol.

To improve on the CCSD description, we go to the next level of coupled-cluster theory, including corrections from triple excitations – see the third row of Table 1.4, where we have listed the triples corrections to the energies as obtained at the CCSD(T) level. The triples corrections to the molecular and atomic energies are almost two orders of magnitude smaller than the singles and doubles corrections. However, for the triples, there is less cancellation between the corrections to the molecule and its atoms than for the doubles. The total triples correction to the AE is therefore only one order of magnitude smaller than the singles and doubles corrections.

With the triples correction added, the error relative to experiment is still as large as 15 kJ/mol. More importantly, we are now above experiment and it is reasonable to assume that the inclusion of higher-order excitations (in particular quadruples) would increase this discrepancy even further, perhaps by a few kJ/mol (judging from the differences between the doubles and triples corrections). Extending the coupled-cluster expansion to infinite order, we would eventually reach the exact solution to the nonrelativistic clamped-nuclei electronic Schrödinger equation, with an error of a little more than 15 kJ/mol. Clearly, for agreement with experiment, we must also take into account the effects of nuclear motion and relativity.

From Table 1.4, we note that the zero-point vibrational energy (ZPVE) correction is large and negative, reducing the error at the CCSD(T) level to only 2.2 kJ/mol. A further inclusion of the first-order relativistic correction brings the error down to only 0.2 kJ/mol, an excellent result. However, before we become too enthusiastic about this result, it should be pointed out that the error in the perturbative treatment of the triples correction in CCSD(T) is quite large (2 kJ/mol, see Table 1.2), partly cancelling the error that arises from the neglect of quadruple and higher excitations. In addition, there are unknown (but probably small) non-Born-Oppenheimer corrections. In conclusion, it seems possible to calculate AEs to an accuracy of 1 - 2 kJ/mol, but very difficult to reduce it further. We shall later present a statistical analysis (based on more molecules) that confirms this tentative conclusion.

Table 1.5 The basis-set dependence of the AE of the CO molecule (kJ/mol). The calculations have been carried out at the all-electron CCSD(T) level at the geometry optimized in the cc-pCVQZ basis [25].

Basis	N_{bas}	HF	SD	(T)	CCSD(T)	Error
cc-pCVDZ	36	710.2	277.4	24.5	1012.1	-74.8
cc-pCVTZ	86	727.1	297.3	32.6	1057.0	-29.9
cc-pCVQZ	168	730.3	311.0	33.8	1075.1	-11.8
cc-pCV5Z	290	730.1	316.4	34.2	1080.7	-6.2
cc-pcV6Z[a]	460	730.1	318.8	34.4	1083.3	-3.6
Limit	∞	730.1	322.1	34.6	1086.9	0.0

[a] Approximate cc-pCV6Z result, see section 6.2.

4.2. Dependence on the AO Basis Set

In our discussion so far, we have used electronic energies that are assumed to represent calculations carried out in an infinite basis of one-particle functions (the basis-set limit). In practice, finite basis sets are used; as we shall see, the truncation of the one-electron basis is a serious problem that may lead to large errors in the calculations.

As seen from Table 1.5, the convergence with respect to X is slow for the correlation contributions to the AE. Even with the largest basis, we have an error of -3.6 kJ/mol, originating almost exclusively from the basis-set truncation of the doubles contribution to the CCSD energy. The slow convergence arises from the orbital approximation (i.e., the expansion of the wavefunction in determinants), leading to a poor description of the short-range correlated motion of the electrons. Noting that as many as 460 AOs are needed for a small diatomic molecule to achieve chemical accuracy, it is clear that this brute-force approach does not represent a widely applicable tool for the calculation of thermochemical data.

There are two possible solutions to this problem. We may either modify our ansatz for the wavefunction, including terms that depend explicitly on the interelectronic coordinates [26-30], or we may take advantage of the smooth convergence of the correlation-consistent basis sets to extrapolate to the basis-set limit [6, 31-39]. In our work, we have considered both approaches; as we shall see, they are fully consistent with each other and with the available experimental data. With these techniques, the accurate calculation of AEs is achieved at a much lower cost than with the brute-force approach described in the present section.

5. SHORT-RANGE CORRELATION AND THE COULOMB HOLE

In the preceding section, we observed the slow basis-set convergence of the doubles contributions to the AE of CO. In the present section, we shall make an attempt at understanding the reasons for the slow convergence and to see if this insight can help us design better computational schemes.

5.1. Terms Linear in r_{12}

The slow convergence of the doubles contributions to the AE is a general problem related to the accurate description of electron pairs in any electronic system. This problem has been studied carefully for the simplest two-electron system, namely the ground-state He atom. In nonrelativistic theory, its Hamiltonian reads

$$\hat{H} = -\frac{1}{2}\nabla_1^2 - \frac{1}{2}\nabla_2^2 - \frac{Z}{r_1} - \frac{Z}{r_2} + \frac{1}{r_{12}}, \tag{5.1}$$

where (in atomic units) the first two and the last three terms represent the kinetic energies of the two electrons and the Coulomb interactions between the three particles, respectively. As two particles coalesce, the potential part of the Hamiltonian becomes singular in the left-hand side of the Schrödinger equation

$$\hat{H}\,\Psi = E\,\Psi. \tag{5.2}$$

For the right-hand side to remain finite, there must be a compensating term arising from the kinetic energy part on the left-hand side. In particular, for the singlet ground state, Slater found that the wavefunction must satisfy the following *cusp conditions* for coalescing particles [40, 41]:

$$\left.\frac{\partial\Psi}{\partial r_i}\right|_{r_i=0} = -Z\,\Psi(r_i=0), \tag{5.3}$$

$$\left.\frac{\partial\Psi}{\partial r_{12}}\right|_{r_{12}=0} = \tfrac{1}{2}\,\Psi(r_{12}=0) \tag{5.4}$$

(spherical averaging implied). These conditions are satisfied if, for example, the wavefunction behaves in the following manner for small interparticle distances:

$$\Psi(r_i \approx 0) = \exp(-Z\,r_i)\,\Psi(r_i=0), \tag{5.5}$$

$$\begin{aligned}\Psi(\mathrm{r}_{12} \approx 0) &= \exp(\tfrac{1}{2}\,\mathrm{r}_{12})\,\Psi(\mathrm{r}_{12} = 0) \\ &= \left[1 + \tfrac{1}{2}\,\mathrm{r}_{12} + \mathcal{O}(\mathrm{r}_{12}^2)\right]\Psi(\mathrm{r}_{12} = 0). \end{aligned} \tag{5.6}$$

Whereas the one-electron exponential form Eq. (5.5) is easily implemented for orbital-based wavefunctions, the explicit inclusion in the wavefunction of the interelectronic distance Eq. (5.6) goes beyond the orbital approximation (the determinant expansion) of standard quantum chemistry since r_{12} does not factorize into one-electron functions. Still, the inclusion of a term in the wavefunction containing r_{12} linearly has a dramatic impact on the ability of the wavefunction to model the electronic structure as two electrons approach each other closely.

To see the importance of the r_{12} term, consider the standard FCI expansion of the He ground-state wavefunction. The FCI wavefunction is written as a linear expansion of determinants,

$$\Psi_{\mathrm{FCI}} = \sum_{\mathrm{K}} \mathrm{C}_{\mathrm{K}}\,\Phi_{\mathrm{K}}, \tag{5.7}$$

each of which contains a product of two Slater-type orbitals (STOs),

$$\chi_{n\ell m}(\mathbf{r}) = \mathrm{r}^{n-1}\exp(-\zeta \mathrm{r})\,\mathrm{Y}_{\ell m}(\theta,\varphi), \tag{5.8}$$

where $\mathrm{Y}_{\ell m}(\theta,\varphi)$ is a spherical-harmonic function. The same (optimized) exponent ζ is used for all STOs, which differ only in the quantum numbers $n > \ell \geq |m| \geq 0$. By including in the FCI wavefunction all STOs up to a given principal quantum number $n = \mathrm{X}$, a sequence of FCI wavefunctions is established, which approaches the exact nonrelativistic wavefunction as X tends to infinity. In the following, we shall refer to this hierarchy of FCI wavefunctions as the *principal expansion* [12]. For $\mathrm{X} = 1$, the principal expansion contains only one determinant (the Hartree-Fock determinant); for $\mathrm{X} > 1$, the FCI wavefunction is a multideterminant expansion.

To illustrate the convergence of the FCI principal expansion with respect to short-range electron correlation, we have in Fig. 1.1 plotted the ground-state He wavefunction with both electrons fixed at a distance of 0.5 a_0 from the nucleus, as a function of the angle θ_{12} between the position vectors $\mathbf{r}_1$ and $\mathbf{r}_2$ of the two electrons. The thick grey lines correspond to the exact nonrelativistic wavefunction, whereas the FCI wavefunctions are plotted using black lines. Clearly, the description of the Coulomb cusp and more generally the Coulomb hole is poor in the orbital approximation. In particular, no matter how many terms we include in the FCI wavefunction, we will not be able to describe the nondifferentiability of the wavefunction at the point of coalescence.

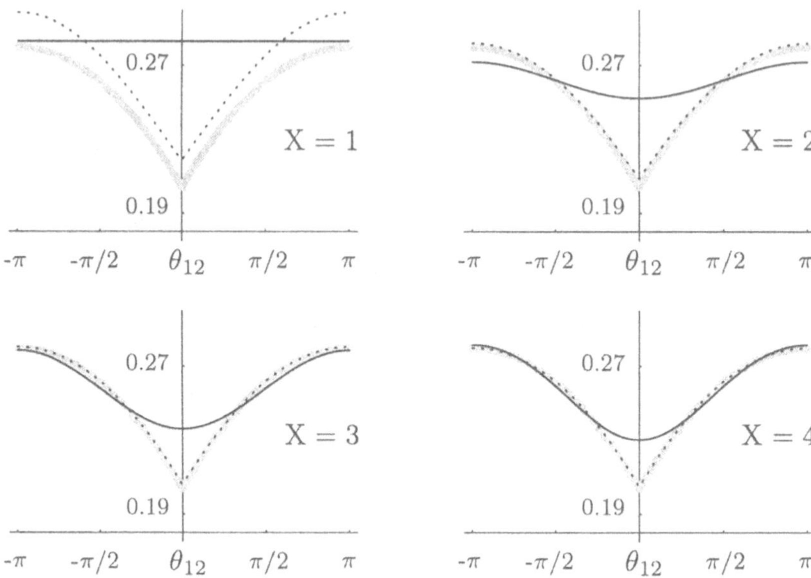

Figure 1.1 The ground-state He wavefunction plotted on a circle of radius 0.5 a_0 about the nucleus with one electron fixed at the origin of the plot. The black lines represent the FCI wavefunction, Eq. (5.7), for $1 \leq X \leq 4$; the dotted lines represent the corresponding r_{12}-FCI wavefunction, Eq. (5.9). The thick grey line corresponds to the exact nonrelativistic wavefunction.

However, this deficiency of the FCI expansion is easily rectified by including in the wavefunction a single extra term that is linear in the interelectronic distance. The resulting wavefunction may be written as [42]

$$\Psi_{r_{12}-\mathrm{FCI}} = \left\{ \sum_{K} C'_K \, \Phi_K + C_R \, r_{12} \exp\left[-\zeta \left(r_1 + r_2\right)\right] \right\} \left(\alpha_1\beta_2 - \alpha_2\beta_1\right), \tag{5.9}$$

where the coefficients of the CI expansion and of the r_{12} term are optimized simultaneously with the orbital exponent ζ. The corresponding wavefunctions are plotted using a dotted line in Fig. 1.1. The improvement in the description of the Coulomb hole is dramatic – already when the r_{12} term is added only to the Hartree-Fock determinant. The improvement in the energy is just as impressive. Whereas the standard FCI principal expansion has errors of 147, 37, 12, and 6.5 kJ/mol for $1 \leq X \leq 4$, the corresponding errors with the r_{12} term included are only 33, 2.3, 0.39, and 0.14 kJ/mol.

The use of the linear r_{12} term for many-electron systems is more involved but has been successfully incorporated in the framework of coupled-cluster theory by Kutzelnigg and coworkers [43-46]. In their R12 theory, the difficult many-electron integrals that arise from the inclusion of the interelectronic distance in the wavefunction are avoided by the resolution-of-identity approximation, yielding a highly efficient scheme for the accurate calculation of atomic and molecular electronic energies. For example, comparing with standard coupled-cluster calculations, the evaluation of the R12 two-electron integrals requires only about four times more computation time; the remaining part of the calculation requires essentially no additional computational effort.

5.2. Extrapolations from Principal Expansions

Although the convergence of the FCI principal expansion is slow, it is systematic [12]. In fact, for a sufficiently large basis, it has been found that each STO in the FCI principal expansion of He contributes an amount of energy that, to a good approximation, is given by the expression [47, 48]

$$\Delta \mathrm{E}_{n\ell m} \propto n^{-6}. \tag{5.10}$$

Note that the energy contribution depends only on the principal quantum number n. Therefore, each of the n^2 orbitals that constitute the shell n contributes the same amount of energy, justifying the use of the principal expansion. Summing the energy contributions from all orbitals, we obtain

$$\Delta \mathrm{E}_n = \sum_{\ell=0}^{n-1} \sum_{m=-\ell}^{\ell} \Delta \mathrm{E}_{n\ell m} \propto n^{-4}. \tag{5.11}$$

By summing the contributions from all neglected shells, it is now easy to estimate the error that arises when the principal expansion is truncated after $n = \mathrm{X}$:

$$\mathrm{E_X} - \mathrm{E}_\infty = - \sum_{n>\mathrm{X}} \Delta \mathrm{E}_n \propto \mathrm{X}^{-3}. \tag{5.12}$$

This empirical result is consistent with the theoretical analysis of the partial-wave expansion (where the truncation of the FCI expansion is based on the angular-momentum quantum number ℓ rather than on the principal quantum number n), for which it has been proved that the truncation error is proportional to L^{-3} when all STOs up to $\ell = \mathrm{L}$ are included in the FCI wavefunction [49, 50].

Guided by Eq. (5.12), we assume that the calculated He energy is well represented by the expression

$$\mathrm{E_X} = \mathrm{E}_\infty + \mathrm{a}\,\mathrm{X}^{-3}. \tag{5.13}$$

This equation contains two unknowns and we can thus extrapolate to the basis-set limit from two separate calculations with different cardinal numbers X and Y. This gives us the following simple expression for the energy at the basis-set limit [32, 33]:

$$E_\infty = \frac{E_X X^3 - E_Y Y^3}{X^3 - Y^3}. \tag{5.14}$$

For example, if we carry out calculations with X = 3 and Y = 4 using optimized numerical orbitals (i.e., no longer simple STOs), we obtain errors in the energy of 4.9 and 2.1 kJ/mol, respectively. The error in the energy extrapolated from these two results using Eq. (5.14) is less than 0.1 kJ/mol, which would require a FCI principal expansion with X = 10 or more.

6. CALIBRATION OF THE EXTRAPOLATION TECHNIQUE

6.1. Valence-Shell Correlation Energy

In section 4, we established that the orbital truncation error represents a serious obstacle to the accurate calculation of AEs. Next, in section 5, we found that this problem may be solved in two different ways: we may either employ wavefunctions that contain the interelectronic distance explicitly (in particular the R12 model), or we may try to extrapolate to the basis-set limit using energies obtained with finite basis sets. In the present section, we shall apply both methods to a set of small molecules, to establish whether or not these techniques are useful also for systems of chemical interest.

It is important to realize that molecular electronic systems differ from the He atom in the sense that the uncorrelated Hartree-Fock description cannot be expressed in terms of a single, doubly occupied $1s$ orbital. In particular, in sequences of calculations using the correlation-consistent orbitals, we observe not only changes related to the improved description of electron correlation, but also changes in the uncorrelated Hartree-Fock description. Within the Hartree-Fock model, effects such as polarization of the atomic charge distributions upon molecular formation require the use of flexible basis sets, albeit convergence is usually reached much more rapidly than for the description of electron correlation. Therefore, in order to study the asymptotic convergence of the short-range correlation problem, we must first subtract from the total electronic energy the Hartree-Fock energy. In passing, we note that the

Table 1.6 CCSD valence correlation energies calculated with the cc-pVXZ basis sets and using the extrapolation formula Eq. (5.14) are compared with the R12 values (mE_h). The last column contains the mean absolute deviations from the R12 energies. All calculations have been carried out at the optimized all-electron CCSD(T)/cc-pCVQZ geometries [25].

	CH_2 [a]	H_2O	HF	N_2	CO	Ne	F_2	$\bar{\Delta}_{abs}$
D	-138.0	-211.2	-206.8	-309.4	-294.5	-189.0	-402.7	107.9
T	-164.2	-267.4	-273.9	-372.0	-358.3	-266.3	-526.0	39.8
Q	-171.4	-286.0	-297.6	-393.2	-380.6	-294.7	-569.7	16.2
5	-173.6	-292.4	-306.3	-400.7	-388.5	-305.5	-586.1	7.7
6	-174.5	-294.9	-309.7	-403.7	-391.7	-309.9	-592.8	4.2
DT	-175.2	-291.1	-302.2	-398.3	-385.1	-298.9	-577.9	11.2
TQ	-176.7	-299.5	-314.9	-408.7	-396.9	-315.4	-601.6	1.0
Q5	-176.0	-299.2	-315.3	-408.6	-396.8	-316.8	-603.4	1.3
56	-175.7	-298.3	-314.4	-407.9	-396.1	-316.0	-601.9	0.5
R12	-175.5	-297.9	-313.9	-407.4	-395.7	-315.5	-601.0	

[a] $^1A_1$ state.

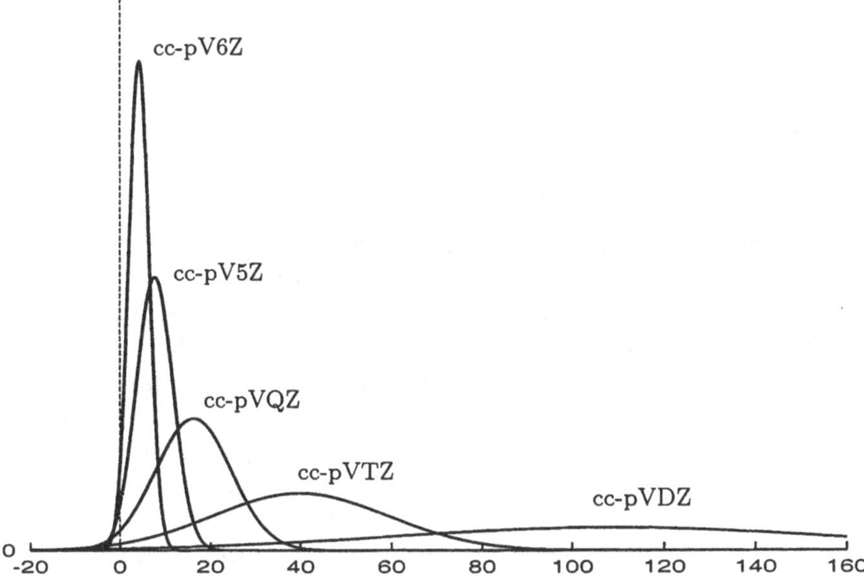

Figure 1.2 Normal distributions of the errors in the calculated valence-shell CCSD/cc-pVXZ correlation energies relative to the R12 reference values (mE_h).

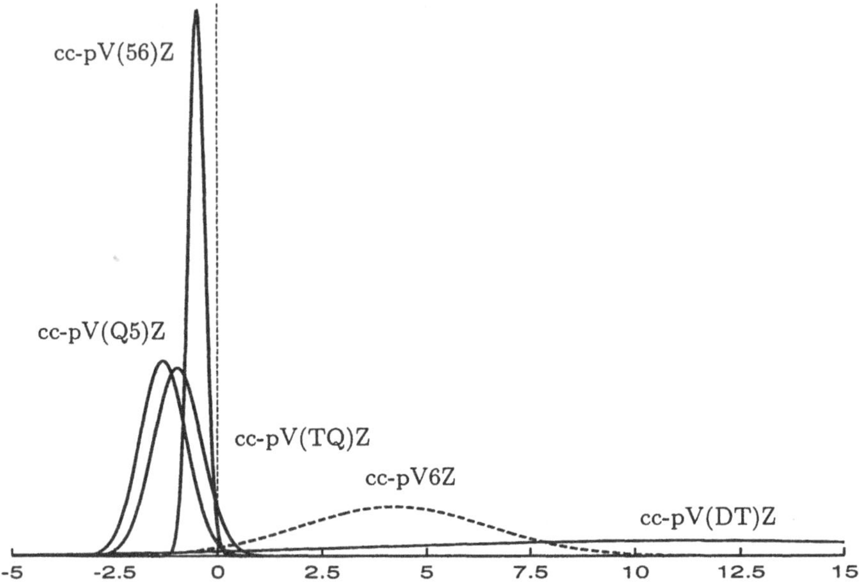

Figure 1.3 Normal distributions (solid lines) of the errors in the extrapolated valence-shell CCSD/cc-pV(X - 1,X)Z correlation energies relative to the R12 reference values (mE_h). The corresponding CCSD/cc-pV6Z distribution (dashed line, see Fig. 1.2) is also shown for comparison.

Hartree-Fock convergence does not present an insurmountable difficulty; there are clear indications that the molecular Hartree-Fock energy converges as $\exp(-\alpha X)$ and thus rather rapidly [51-53].

Table 1.6 displays the CCSD valence correlation energies of six small molecules and the Ne atom in units of 1 mE_h = 2.6255 kJ/mol. The first five rows contain the energies calculated in the standard manner in the cc-pVXZ basis sets with $2 \leq X \leq 6$. As expected, the convergence is slow, with errors relative to the R12 energies (contained in the last row [54]) of about 10 kJ/mol or more, even for the largest basis sets. This can also be seen from Fig. 1.2, which shows how the calculated CCSD/cc-pVXZ energies converge slowly but smoothly towards the R12 valence-shell correlation energies.

This convergence is significantly accelerated by applying the extrapolation formula (5.14), the errors being reduced to 1 - 3 kJ/mol for all extrapolations except for the cc-pV(DT)Z energies. (Here and elsewhere we shall use the notation cc-pV(X - 1,X)Z for the energy obtained by extrapolation from the cc-pV(X - 1)Z and cc-pVXZ correlation energies.)

Table 1.7 CCSD(T) total energies calculated with the cc-pCVXZ basis sets and compared with the corresponding experimental total energies (E_h). The last row contains the mean absolute deviations from the experimental energies. All calculations have been carried out at the optimized all-electron CCSD(T)/cc-pCVQZ geometries [25].

	D	T	Q	5	6[a]	Exp.[b]
CH_2 [c]	-39.060	-39.111	-39.126	-39.130	-39.131	-39.133
H_2O	-76.282	-76.390	-76.421	-76.431	-76.434	-76.439
HF	-100.270	-100.400	-100.438	-100.450	-100.454	-100.460
N_2	-109.355	-109.482	-109.520	-109.532	-109.535	-109.543
CO	-113.134	-113.264	-113.303	-113.315	-113.319	-113.326
Ne	-128.722	-128.868	-128.912	-128.927	-128.932	-128.938
F_2	-199.182	-199.419	-199.488	-199.511	-199.519	-199.531
$\bar{\Delta}_{abs}$	0.1948	0.0622	0.0231	0.0106	0.0066	

[a] cc-pcV6Z results, see text.
[b] Obtained from experimental dissociation energies and estimated total atomic energies, see text.
[c] $^1A_1$ state.

Fig. 1.3 shows the normal distributions of the errors of the cc-pV(X - 1,X)Z extrapolations (note that the scale is different from Fig. 1.2). In comparison with the cc-pV6Z results, which are included in Fig. 1.3 as a broad distribution on its scale, the agreement between the R12 energies and the extrapolated energies is excellent, confirming our earlier conclusions from the discussion of the He atom.

6.2. Total Electronic Energy

Having observed the agreement between R12 and extrapolation for molecular systems, let us now compare directly with experiment. In Table 1.7, we compare the all-electron CCSD(T)/cc-pCVXZ energies of the seven systems with a set of empirically estimated nonrelativistic total electronic energies, obtained by combining the atomic energies compiled by Chakravorty and Davidson [55] with the experimental equilibrium AEs of Bak et al. [9]. The sextuple-zeta results were obtained from the valence-electron cc-pV6Z energies by adding the differences between the all-electron cc-pCV5Z and valence-electron cc-pV5Z energies. The resulting energies are denoted cc-pcV6Z and should be close to the true cc-pCV6Z energies.

As expected from our previous discussion of the CCSD valence correlation energies, the convergence towards the experimental energies is slow, with mean absolute errors of 511, 163, 61, 28, and 17 kJ/mol as

Table 1.8 CCSD(T) total energies calculated using the R12 method as well as the extrapolation formula (5.14) with the basis sets cc-pCVXZ, compared with the corresponding experimental total energies (E_h). The last row contains the mean absolute deviations from the experimental energies. All calculations have been carried out at the optimized all-electron CCSD(T) /cc-pCVQZ geometries [25].

	R12	DT	TQ	Q5	56[a]	Exp.[b]
CH_2 [c]	-39.133	-39.128	-39.134	-39.134	-39.133	-39.133
H_2O	-76.438	-76.422	-76.438	-76.439	-76.439	-76.439
HF	-100.459	-100.438	-100.459	-100.461	-100.460	-100.460
N_2	-109.540	-109.524	-109.542	-109.542	-109.542	-109.543
CO	-113.325	-113.306	-113.326	-113.326	-113.326	-113.326
Ne	-128.938	-128.911	-128.936	-128.939	-128.938	-128.938
F_2	-199.528	-199.491	-199.527	-199.530	-199.529	-199.531
$\bar{\Delta}_{abs}$	0.0014	0.0214	0.0014	0.0004	0.0005	

[a] In the extrapolation, we have used the cc-pcV6Z results, see text.
[b] Obtained from experimental dissociation energies and estimated total atomic energies, see text.
[c] $^1A_1$ state.

we go from cc-pCVDZ to cc-pcV6Z. In particular, even at the cc-pcV6Z level, the mean absolute error of 17 kJ/mol is significantly larger than the intrinsic error of the CCSD(T) model. Without the benefit of a systematic cancellation of errors, calculations carried out with these basis sets would not be sufficiently accurate.

Let us now compare the R12 energies [54] and the extrapolated energies with the experimental energies for the same systems; see Table 1.8. The R12 energies are between 2 and 4 kJ/mol above the experimental estimates – a significant improvement over the cc-pcV6Z results in Table 1.7. Clearly, the precision of the R12 method is sufficiently high to make the intrinsic error of the CCSD(T) model become an important consideration. Moreover, heats of reaction can be calculated accurately without having to rely on a large cancellation of errors among the products and reactants.

Turning our attention to the extrapolated CCSD(T) energies, we find that the accuracy of the R12 method is attained already at the cc-pCV(TQ)Z level, although the standard deviation in the cc-pCV(TQ)Z errors is somewhat larger than for the R12 errors. The cc-pCV(DT)Z energies are considerably less accurate, but still as good as the cc-pVQZ energies at a much reduced cost. Finally, with a mean absolute deviation of 1 kJ/mol and a maximum deviation of 5 kJ/mol, the cc-pCV(Q5)Z and cc-pcV(56)Z energies agree with their experimental counterparts,

Table 1.9 Calculated all-electron CCSD(T) equilibrium AEs compared with experiment (kJ/mol). All calculations have been carried out at the optimized all-electron CCSD(T)/cc-pCVQZ geometries [25].

	CH_2 [a]	H_2O	HF	N_2	CO	F_2	$\bar{\Delta}_{abs}$
D	692.8	875.3	529.6	841.2	1013.4	112.2	77.9
T	739.3	943.3	573.6	911.8	1058.6	146.7	26.4
Q	751.3	963.5	586.1	936.3	1075.5	153.4	10.9
5	754.9	970.5	590.4	945.6	1080.9	156.9	5.4
6	756.2	972.7	591.6	949.6	1083.2	158.8	3.3
DT	753.3	961.4	583.6	930.1	1070.3	154.4	13.1
TQ	758.7	974.7	592.8	952.3	1086.6	159.5	1.7
Q5	758.0	976.2	593.8	954.9	1087.0	160.9	1.1
56	757.9	975.5	593.3	954.9	1086.9	161.1	0.8
Exp. [b]	757.1	975.3	593.2	956.3	1086.9	163.4	

[a] $^1A_1$ state.
[b] From Ref. 9.

making these levels of theory well suited for computational thermochemistry.

Although the agreement between calculated and experimental correlation and total energies is reassuring, as chemists we are more interested in relative quantities. Let us therefore turn our attention to AEs. In Table 1.9, we compare the calculated all-electron CCSD(T) equilibrium AEs with the corresponding AEs derived from experimental data, see Ref. 9.

Without extrapolation, the errors are reduced by a factor of two to three compared with the errors in the total electronic energies, reflecting the systematic nature of the errors in the total energies. For low cardinal numbers, the same is true for the extrapolated AEs. However, for the cc-pCV(Q5)Z and cc-pcV(56)Z data, the errors in the AEs are similar to the errors in the corresponding total energies, indicating the presence of statistical errors of the order of 1 kJ/mol in the experimental and extrapolated energies. From a practical point of view, we note that the cc-pCV(TQ)Z AEs agree with their experimental counterparts to within 2 kJ/mol, suggesting that chemical accuracy in calculated AEs and heats of reaction should be obtainable at the all-electron CCSD(T)/cc-pCV(TQ)Z level of theory.

6.3. Core Contributions to AEs

In the calculations presented so far, all electrons have been correlated. However, chemical reactions involve mainly the valence electrons, leaving the core electrons nearly unaffected. It is therefore tempting to correlate only the valence electrons and to let the core orbitals remain doubly occupied. In this way, we avoid the calculation of the nearly constant core-correlation energy, concentrating on the valence correlation energy. The freezing of the core electrons simplifies the calculations as there are fewer electrons to correlate and since it enables us to use the cc-pVXZ basis sets rather than the larger cc-pCVXZ sets.

Nevertheless, core-correlation contributions to AEs are often sizeable, with contributions of about 10 kJ/mol for some of the molecules considered here (CH_4, C_2H_2, and C_2H_4). For an accuracy of 10 kJ/mol or better, it is therefore necessary to make an estimate of core correlation [9, 56]. It is, however, not necessary to calculate the core correlation at the same level of theory as the valence correlation energy. We may, for example, estimate the core-correlation energy by extrapolating the difference between all-electron and valence-electron CCSD(T) calculations in the cc-pCVDZ and cc-pCVTZ basis sets. The core-correlation energies obtained in this way reproduce the CCSD(T)/cc-pCV(Q5)Z core-correlation contributions to the AEs well, with mean absolute and maximum deviations of only 0.4 kJ/mol and 1.4 kJ/mol, respectively. By contrast, the calculation of the valence contribution to the AEs by cc-pCV(DT)Z extrapolation leads to errors as large as 30 kJ/mol.

7. MOLECULAR VIBRATIONAL CORRECTIONS

The total energy E_0 of a molecular system in its vibrational ground state can be written as the sum of the electronic energy E_e at the equilibrium geometry and the zero-point vibrational energy (ZPVE), denoted as E_{ZPV},

$$E_0 = E_e + E_{ZPV}. \tag{7.1}$$

The ZPVE may be partitioned into harmonic and anharmonic contributions

$$\begin{aligned} E_{ZPV} &= E_{ZPV}^{hrm} + E_{ZPV}^{anh} \\ &= \frac{1}{2} \sum_i d_i \, \omega_i + \frac{1}{4} \sum_{i \geq j} d_i \, d_j \, X_{ij}, \end{aligned} \tag{7.2}$$

Table 1.10 Harmonic and anharmonic ZPVE contributions, and first-order relativistic contributions to AEs (kJ/mol).

	ZPVE			
	Harm.	Anh.	MVD[a]	SO
CH_2 [b]	-43.22[c]	n.a.[d]	-0.37	-0.35
CO	-12.98	0.04	-0.68	-1.29
F_2	-5.48	0.03	-0.12	-3.22
H_2O	-56.36	0.92	-1.14	-0.93
HF	-24.75	0.27	-0.84	-1.61
N_2	-14.11	0.04	-0.58	0.00

[a] MV + 1D calculated at the CCSD(T)/cc-pCVQZ level.

[b] $^1A_1$ state.

[c] Total ZPVE.

[d] Not available.

where d_i is the degeneracy of the vibrational mode i , ω_i is the harmonic frequency, and X_{ij} are the anharmonic constants.

The harmonic frequencies and the anharmonic constants may be obtained from experimental vibrational spectra, although their determination becomes difficult as the size of the system increases. In Table 1.10, we have listed experimental harmonic and anharmonic contributions to the AEs. These contributions may also be obtained from electronic-structure calculations of quadratic force fields (for harmonic frequencies) and cubic and quartic force fields (for anharmonic constants). For some of the larger molecules in Table 1.11, we have used ZPVEs calculated at the CCSD(T)/cc-pVTZ level or higher, see Ref. 12. In some cases, both experimental and theoretical ZPVEs are available and agree to within 0.3 kJ/mol [12, 57].

Although the harmonic ZPVE must always be taken into account in the calculation of AEs, the anharmonic contribution is much smaller (but oppositely directed) and may sometimes be neglected. However, for molecules such as H_2O, NH_3, and CH_4, the anharmonic corrections to the AEs amount to 0.9, 1.5, and 2.3 kJ/mol and thus cannot be neglected in high-precision calculations of thermochemical data. Comparing the harmonic and anharmonic contributions, it is clear that a treatment that goes beyond second order in perturbation theory is not necessary as it would give contributions that are small compared with the errors in the electronic-structure calculations.

8. RELATIVISTIC CONTRIBUTIONS

Up to this point, we have considered the nonrelativistic Schrödinger equation. However, to calculate AEs to an accuracy of a few kJ/mol, it is necessary to account for relativistic effects, even for molecules containing only hydrogen and first-row atoms. Fortunately, the major relativistic contributions to the AEs of such molecules – the mass-velocity (MV), one-electron Darwin (1D), and first-order spin-orbit (SO) terms – are easily obtained [58].

Whereas the SO corrections are accurately known from atomic measurements, the MV and 1D corrections must be calculated as the expectation values of the operators

$$\hat{\mathrm{H}}_{\mathrm{MV}} = -\frac{1}{8\mathrm{c}^2}\sum_{\mathrm{i}} \nabla_{\mathrm{i}}^4; \qquad \hat{\mathrm{H}}_{1\mathrm{D}} = \frac{\pi}{2\mathrm{c}^2}\sum_{\mathrm{iI}} \mathrm{Z}_{\mathrm{I}}\,\delta(\mathbf{r}_{\mathrm{iI}}), \tag{8.1}$$

where the summations are over all electrons and nuclei and where $\mathrm{c} = 137.036\ \mathrm{E_h a_0/\hbar}$ is the velocity of light. Since the MV operator is a correction to the kinetic-energy operator and the 1D operator a correction to the nuclear-attraction operator involving Dirac delta functions at the nuclei, it is evident that their expectation values depend primarily on the core of the electronic wavefunction. As for the total nonrelativistic energy (see Table 1.4), we therefore expect large cancellations of the contributions from these terms to the AEs. As seen from Table 1.10, the MV and 1D corrections to the AEs are of the order of 1 kJ/mol; the corrections to the total electronic energies are three orders of magnitude larger. The SO corrections to the total energies are much smaller than the scalar corrections but do not cancel since they occur only for the atoms. Therefore, the total relativistic corrections to the AEs amount to a few kJ/mol and must be taken into account in calculations at the CCSD(T) level.

Our relativistic treatment is incomplete in the sense that only first-order corrections are considered. For systems containing only first-row elements, the higher-order corrections are small and may be safely neglected for a target accuracy of 1 kJ/mol. For higher accuracy, we would also have to include a number of nonrelativistic corrections such as the mass-polarization and diagonal adiabatic non-Born-Oppenheimer corrections [59]. Since the underlying CCSD(T) model is anyway incapable of such a high precision, we ignore such corrections here.

9. CALCULATION OF ATOMIZATION ENERGIES

In Table 1.11, the AEs are listed for twenty small molecules. The AEs are obtained by adding vibrational and relativistic corrections to the nonrelativistic CCSD(T)/cc-pcV(56)Z equilibrium AEs. The ZPVEs have been taken from the compilation of Helgaker, Jørgensen, and Olsen [12]; the relativistic contributions contain the MV and 1D scalar corrections calculated at the CCSD(T)/cc-pCVQZ level, in addition to first-order SO corrections from atomic measurements [9]. Table 1.11 also contains experimental AEs.

The calculated AEs are very accurate, with typical errors of about 1 kJ/mol and errors larger than 2.3 kJ/mol occurring only for O_3 (-10.7 kJ/mol) and HOF (-12.0 kJ/mol). For O_3, the difference probably arises from an error in the calculation as the high accuracy of the CCSD(T) model does not extend to systems that are poorly represented by a single determinant. For the single-determinant HOF molecule, the discrepancy is most likely caused by an error in the tabulated value derived from experimental data.

The CCSD(T)/cc-pcV(56)Z calculations are computationally demanding and can be carried out only for small molecules. In Table 1.12, we compile the statistical errors that have been obtained with smaller basis sets. In compiling the statistics, we have excluded HOF (since the experimental value is in doubt) and O_3 so as to obtain errors typical of single-determinant molecules.

Our statistical analysis reveals a large improvement from cc-pCV(DT)Z to cc-pCV(TQ)Z; see Fig. 1.4. In fact, the cc-pCV(TQ)Z calculations are clearly more accurate than their much more expensive cc-pcV6Z counterparts and nearly as accurate as the cc-pcV(56)Z extrapolations. The cc-pCV(TQ)Z extrapolations yield mean and maximum absolute errors of 1.7 and 4.0 kJ/mol, respectively, compared with those of 0.8 and 2.3 kJ/mol at the cc-pcV(56)Z level. Chemical accuracy is thus obtained at the cc-pCV(TQ)Z level, greatly expanding the range of molecules for which ab initio electronic-structure calculations will afford thermochemical data of chemical accuracy.

10. CONCLUSIONS AND PERSPECTIVES

Quantum chemistry has reached the stage where it is possible to calculate gas-phase thermochemical data to a precision of a few kJ/mol. Although computationally expensive, such calculations can be carried out routinely for a broad range of molecules containing first-row atoms.

Table 1.11 All-electron CCSD(T)/cc-pcV(56)Z equilibrium AEs, molecular zero-point vibrational energies (ZPVE), and relativistic corrections (Rel.), adding up to the calculated AEs, which are compared with the experimental AEs. All calculations have been performed at the optimized all-electron CCSD(T)/cc-pCVQZ geometries [25].

	CCSD(T)	ZPVE	Rel.	Calc. AE	Exp. AE
F_2	161.1	-5.5	-3.3	152.3	154.6±0.6
H_2	458.1	-26.0	-0.0	432.2	432.1±0.01
HF	593.3	-24.5	-2.5	566.3	566.2±0.7
O_3	605.5	-17.4	-3.9	584.3	595.0±1.7
HOF	662.9	-35.9	-3.5	623.5	635.5±4.2
CH_2 [a]	757.9	-43.2	-0.7	714.0	713.1±2.2
HNO	860.4	-35.8	-2.1	822.5	823.6±0.3
N_2	954.9	-14.1	-0.6	940.3	941.6±0.2
H_2O	975.5	-55.4	-2.1	918.0	917.8±0.2
CO	1086.9	-12.9	-2.0	1072.0	1071.8±0.5
H_2O_2	1126.1	-68.0	-3.5	1054.6	1055.5
HNC	1247.8	-40.6	-1.4	1205.8	n.a.[b]
NH_3	1247.4	-89.0	-1.1	1157.4	1157.8±0.4
HCN	1311.0	-41.6	-1.3	1268.1	1269.8±2.6
CH_2O	1568.0	-69.1	-2.7	1496.2	1494.7±0.7
CO_2	1633.2	-30.3	-4.2	1598.7	1597.9±0.5
C_2H_2	1697.1	-68.8	-1.9	1626.4	1627.2±1.0
CH_4	1759.4	-115.9	-1.2	1642.3	1642.2±0.6
C_2H_4	2360.8	-132.2	-2.1	2226.5	2225.5±0.7

[a] $^1A_1$ state.

[b] Not available.

The prerequisites for high accuracy are coupled-cluster calculations with the inclusion of connected triples [e.g., CCSD(T)], either in conjunction with R12 theory or with correlation-consistent basis sets of at least quadruple-zeta quality followed by extrapolation. In addition, harmonic vibrational corrections must always be included. For small molecules, such as those contained in Table 1.11, such calculations have errors of the order of a few kJ/mol. To reduce the error below 1 kJ/mol, connected quadruples must be taken into account, together with anharmonic vibrational and first-order relativistic corrections. In practice, the approximate treatment of connected triples in the CCSD(T) model introduces an error (relative to CCSDT) that often tends to cancel the

Table 1.12 Statistical measures of errors for extrapolated CCSD(T) AEs relative to experiment (kJ/mol). $\bar{\Delta}$ is the mean error, Δ_{std} is the standard deviation around the mean error, $\bar{\Delta}_{abs}$ is the mean absolute error, and Δ_{max} is the maximum absolute error.

	DT	TQ	Q5	56
$\bar{\Delta}$	-14.67	-0.23	0.08	-0.44
Δ_{std}	8.39	2.16	1.12	1.05
$\bar{\Delta}_{abs}$	14.74	1.68	0.90	0.84
Δ_{max}	29.53	4.01	2.45	2.31

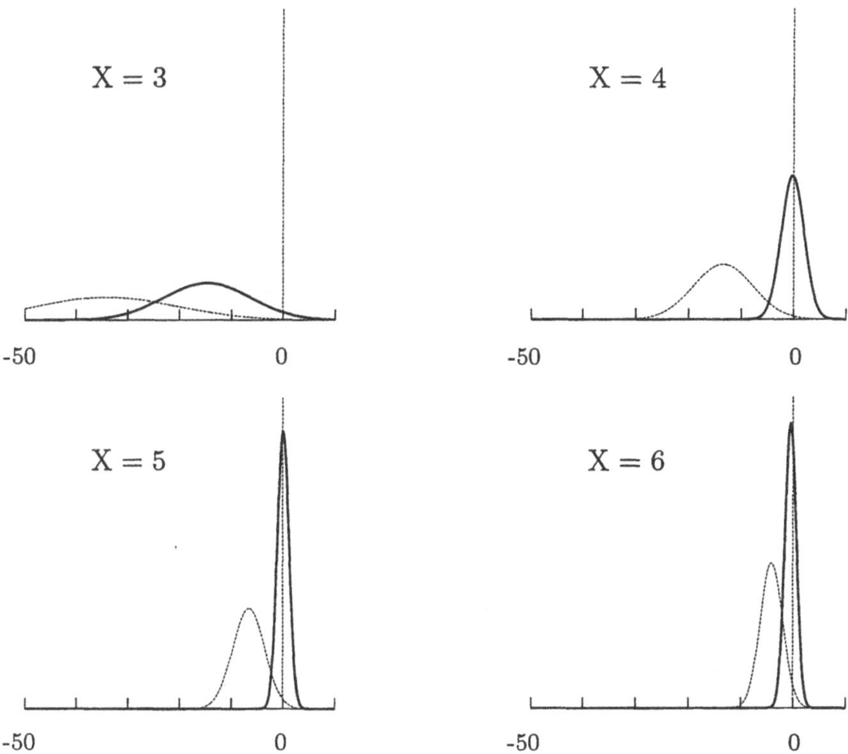

Figure 1.4 Normal distributions of the errors in the calculated cc-pCVXZ AEs (dashed lines) and extrapolated cc-pCV(X - 1,X)Z AEs (solid lines) relative to experiment (kJ/mol).

error arising from the neglect of connected quadruples. Because of this cancellation, CCSD(T) calculations benefit from anharmonic and relativistic corrections, yielding highly accurate results as demonstrated in Table 1.11.

Our study has been restricted to molecules containing only first-row atoms and with wavefunctions dominated by one determinant. Molecules such as O_3 are less accurately described, with an error of about 10 kJ/mol at the CCSD(T) level of theory. For such multiconfigurational systems, more elaborate treatments are necessary and no programs are yet available for routine applications. As we go down the periodic table, relativistic effects become more important and the electronic structures more complicated. Therefore, for such systems it is presently not possible to calculate thermochemical data to the same accuracy as for closed-shell molecules containing first-row atoms. Nevertheless, systems with wavefunctions dominated by single determinant are by far the most abundant and it is promising that the accuracy of a few kJ/mol is obtainable for them.

ACKNOWLEDGEMENTS

We gratefully acknowledge grants of computing time by the Academic Computing Services Amsterdam (SARA) and by the Research Council of Norway (NFR Supercomputing Grant No. NN1118K). A.H. acknowledges support from the European Commission, Marie Curie Individual Fellowships Programme, Contract No. MCFI-1999-00006. The research of W.K. has been made possible by a fellowship of the Royal Netherlands Academy of Arts and Sciences.

REFERENCES

1. J. M. L. Martin, *Chem. Phys. Lett.* **259**, 669 (1996).
2. J. M. L. Martin and P. R. Taylor, *J. Chem. Phys.* **106**, 8620 (1997).
3. G. A. Petersson, D. K. Malick, W. G. Wilson, J. W. Ochterski, J. A. Montgomery, Jr., and M. J. Frisch, *J. Chem. Phys.* **109**, 10570 (1998).
4. D. Feller and K. A. Peterson, *J. Chem. Phys.* **108**, 154 (1998).
5. D. Feller and K. A. Peterson, *J. Chem. Phys.* **110**, 8384 (1998).
6. J. M. L. Martin and G. de Oliveira, *J. Chem. Phys.* **111**, 1843 (1999).
7. W. Klopper, K. L. Bak, P. Jørgensen, J. Olsen, and T. Helgaker, *J. Phys.* **B 32**, R103 (1999).

8. L. A. Curtiss, K. Raghavachari, P. C. Redfern, and J. A. Pople, *J. Chem. Phys.* **112**, 1125 (2000).
9. K. L. Bak, P. Jørgensen, J. Olsen, T. Helgaker, and W. Klopper, *J. Chem. Phys.* **112**, 9229 (2000).
10. T. H. Dunning, Jr., *J. Phys. Chem.* **A 104**, 9062 (2000).
11. A. G. Császár, W. D. Allen, and H. F. Schaefer III, *J. Chem. Phys.* **108**, 9751 (1998).
12. T. Helgaker, P. Jørgensen, and J. Olsen, *Molecular Electronic-Structure Theory*, Wiley, Chichester (2000).
13. R. J. Bartlett, in *Modern Electronic Structure Theory, Part II*, D. R. Yarkony (ed.), World Scientific, Singapore (1995).
14. J. Olsen, P. Jørgensen, T. Helgaker, and O. Christiansen, *J. Chem. Phys.* **112**, 9736 (2000).
15. T. H. Dunning, Jr., *J. Chem. Phys.* **90**, 1007 (1989).
16. D. E. Woon and T. H. Dunning, Jr., *J. Chem. Phys.* **103**, 4572 (1995).
17. A. K. Wilson, T. van Mourik, and T. H. Dunning, Jr., *J. Mol. Struct. (Theochem)* **388**, 339 (1996).
18. J. Olsen, *J. Chem. Phys.* **113**, 7140 (2000).
19. K. Raghavachari, G. W. Trucks, J. A. Pople, and M. Head-Gordon, *Chem. Phys. Lett.* **157**, 479 (1989).
20. K. L. Bak, P. Jørgensen, J. Olsen, T. Helgaker, and J. Gauss, *Chem. Phys. Lett.* **317**, 116 (2000).
21. D. Feller and J. A. Sordo, *J. Chem. Phys.* **112**, 5604 (2000).
22. D. Feller and J. A. Sordo, *J. Chem. Phys.* **113**, 485 (2000).
23. K. P. Huber and G. Herzberg, *Molecular Spectra and Molecular Structure, Vol. IV, Constants of Diatomic Molecules*, Van Nostrand Reinhold, New York (1979).
24. M. W. Chase, Jr. (ed.), *NIST-JANAF Thermochemical Tables, 4th edition*, *J. Phys. Chem.* Ref. Data Monograph, No. 9, American Chemical Society and American Institute of Physics, Woodbury, N.Y. (1998).
25. K. L. Bak, J. Gauss, P. Jørgensen, J. Olsen, T. Helgaker, and J. F. Stanton, *J. Chem. Phys.*, submitted.
26. E. A. Hylleraas, *Z. Phys.* **54**, 347 (1929).
27. W. Cencek and J. Rychlewski, *J. Chem. Phys.* **98**, 1252 (1993).
28. J. Rychlewski, *Adv. Quantum Chem.* **31**, 173 (1998).
29. W. Klopper, in *The Encyclopedia of Computational Chemistry*, P. v. R. Schleyer, N. L. Allinger, T. Clark, J. Gasteiger, P. A. Kollman, H. F. Schaefer III, and P. R. Schreiner (eds.), Wiley, Chichester (1998).
30. R. Bukowski, B. Jeziorski, and K. Szalewicz, *J. Chem. Phys.* **110**, 4165 (1999).
31. D. Feller, *J. Chem. Phys.* **96**, 6104 (1992).
32. T. Helgaker, W. Klopper, H. Koch, and J. Noga, *J. Chem. Phys.* **106**, 9639 (1997).
33. A. Halkier, T. Helgaker, P. Jørgensen, W. Klopper, H. Koch, J. Olsen, and A. K. Wilson, *Chem. Phys. Lett.* **286**, 243 (1998).

34. T. H. Dunning, Jr., K. A. Peterson, and D. E. Woon, in *The Encyclopedia of Computational Chemistry*, P. v. R. Schleyer, N. L. Allinger, T. Clark, J. Gasteiger, P. A. Kollman, H. F. Schaefer III, and P. R. Schreiner (eds.), Wiley, Chichester (1998).
35. P. L. Fast, M. L. Sánchez, and D. G. Truhlar, *J. Chem. Phys.* **111**, 2921 (1999).
36. J. S. Lee and S. Y. Park, *J. Chem. Phys.* **112**, 10746 (2000).
37. R. J. Gdanitz, *J. Chem. Phys.* **113**, 5145 (2000).
38. A. J. C. Varandas, *J. Chem. Phys.* **113**, 8880 (2000).
39. K. L. Bak, A. Halkier, P. Jørgensen, J. Olsen, T. Helgaker, and W. Klopper, *J. Mol. Struct.*, in press.
40. J. C. Slater, *Phys. Rev.* **31**, 333 (1928).
41. T. Kato, *Commun. Pure Appl. Math.* **10**, 151 (1957).
42. W. Kutzelnigg, *Theor. Chim. Acta* **68**, 445 (1985).
43. W. Klopper and W. Kutzelnigg, *Chem. Phys. Lett.* **134**, 17 (1987).
44. W. Kutzelnigg and W. Klopper, *J. Chem. Phys.* **94**, 1985 (1991).
45. J. Noga and W. Kutzelnigg, *J. Chem. Phys.* **101**, 7738 (1994).
46. J. Noga, W. Klopper, and W. Kutzelnigg, in *Recent Advances in Computational Chemistry, Vol. 3, Recent Advances in Coupled-Cluster Methods*, R. J. Bartlett (ed.), World Scientific, Singapore (1997).
47. C. F. Bunge, *Theor. Chim. Acta* **16**, 126 (1970).
48. D. P. Carroll, H. J. Silverstone, and R. M. Metzger, *J. Chem. Phys.* **71**, 4142 (1979).
49. C. Schwartz, *Phys. Rev.* **126**, 1015 (1962).
50. R. N. Hill, *J. Chem. Phys.* **83**, 1173 (1985).
51. A. Halkier, T. Helgaker, P. Jørgensen, W. Klopper, and J. Olsen, *Chem. Phys. Lett.* **302**, 437 (1999).
52. F. Jensen, *J. Chem. Phys.* **110**, 6601 (1999).
53. F. Pahl, Ph.D. thesis, University of Cambridge, UK (1999).
54. W. Klopper, *Mol. Phys.*, in press.
55. S. J. Chakravorty and E. R. Davidson, *J. Phys. Chem.* **A 100**, 6167 (1996).
56. J. M. L. Martin, A. Sundermann, P. L. Fast, and D. G. Truhlar, *J. Chem. Phys.* **113**, 1348 (2000).
57. T. J. Lee, J. M. L. Martin, and P. R. Taylor, *J. Chem. Phys.* **102**, 254 (1995).
58. P. Pyykkö, *Chem. Rev.* **88**, 563 (1988).
59. E. E. Nikitin, *Annu. Rev. Phys. Chem.* **50**, 1 (1999).

Chapter 2

W1 and W2 Theories, and Their Variants: Thermochemistry in the kJ/mol Accuracy Range

Jan M.L. Martin and S. Parthiban
Department of Organic Chemistry, Weizmann Institute of Science, Kimmelman Building, IL-76100 Reḥovot, Israel

1. INTRODUCTION AND BACKGROUND

The last fifteen years witnessed the development of a number of "black-box" computational thermochemistry methods. Among them, the G1/G2/G3 theories and their variants, and the CBS-Q family of methods by Petersson and coworkers are worth mentioning in particular. In addition to these wavefunction-based approaches, density functional methods – aside from their great popularity as a general tool for practical computational chemistry – have gained some currency for computational thermochemistry in the medium accuracy range, as have group equivalent-based models. For very large systems, semiempirical methods remain popular.

At the other extreme in terms of system size and accuracy stand brute-force approaches such as those based on wavefunctions with explicit interelectronic distances.

Methods such as G3 and CBS-QB3 do reach the goal of "chemical accuracy" (generally defined as ± 1 kcal/mol) on average, but worst-case errors for problematic molecules may exceed this criterion by almost an order of magnitude. In addition, almost all of these approaches involve some level of parameterization and/or empirical correction against experimental data. While this is by and large possible (albeit not without pitfalls) in the kcal/mol accuracy range for first-and second-row compounds, experimental data of sub-kcal/mol accuracy are thin on the

J. Cioslowski (ed.), Quantum-Mechanical Prediction of Thermochemical Data, 31–65.

ground, and the available data for transition metal compounds are simply too scarce for this to be a useful approach.

There would thus appear to be room for a more or less "black box" computational thermochemistry method that has the following properties:

1. it on average achieves "benchmark accuracy", which we shall arbitrarily define as one unit of the most common tabulation unit in thermochemical reference tables, i.e. 1 kJ/mol (0.24 kcal/mol);
2. the worst-case error should not exceed 1 kcal/mol ("chemical accuracy") except perhaps in intrinsically pathological cases;
3. it is still efficient enough for applications to systems with up to six heavy atoms on modern workstations;
4. it is entirely devoid of parameters derived from experiment (and hence from bias towards the systems used for parameterization).

These have been the design goals in our development of the W1 and W2 (Weizmann-1 and Weizmann-2) theories [1].

The usual design philosophy for this type of methods is bottom-up: one starts with an approximate model, compares results with experiments, analyzes the deviations, and uses them to determine empirical corrections and/or additional terms to be added to the model, after which the cycle is repeated if desired.

Our philosophy was instead "top-down". We decomposed the molecular TAE (total atomization energy: TAE_e at the bottom of the well, TAE_0 at absolute zero) into all components that can reasonably affect it at the kJ/mol level. Then we carried out exhaustive benchmark calculations on each component *separately* for a representative "training set" of molecules. Finally, for each component separately, we progressively introduced approximations up to the point where reproduction of that particular component started deteriorating to an unacceptable extent. Thus, experimental data entered the picture only at the validation stage, not at the design stage.

Another philosophical issue centers on whether a method should be a "protocol" specified down to the last detail (i.e. be truly "black-box"), or whether it should merely outline a general approach with minor details to be decided on a case-by-case basis. Obviously a method where empirical parameterization is kept to the absolute minimum or is absent altogether will offer more 'degrees of freedom' in this regard than the one where a minor change in the protocol would, for consistency, require reparameterization against a large experimental data set. Yet our general guideline was that, while such choices should be *possible* for an experienced computational chemist, they should not be an essential part of the process itself.

2. STEPS IN THE W1 AND W2 THEORIES, AND THEIR JUSTIFICATION

The more cost-effective W1 theory and the more rigorous W2 theory have a lot of points in common. Aside from issues relating to the reference geometry and the zero-point energy, the main difference concerns the basis sets used in the extrapolation steps for the SCF and the valence correlation contribution.

These basis sets belong to the "correlation consistent" family of Dunning and coworkers [2, 3]. The correlation consistent (cc) basis sets, besides being arguably the most compact ones in their accuracy range [4], have the important property that, by design, they treat radial and angular correlation in a balanced way. In addition to the regular cc-pVnZ (correlation consistent polarized valence n-tuple zeta, or VnZ for short) basis sets, several variants have been published. In particular we note the aug-cc-pVnZ or AVnZ basis sets [5] for anions (with the combination of regular cc-pVnZ on hydrogen and aug-cc-pVnZ on other elements generally being denoted aug′-cc-pVnZ [6], or A′VnZ for short), the MT (Martin-Taylor [7, 8]) and cc-pCVnZ [9] basis sets for inner-shell correlation, and the cc-pVnZ+1 [10], cc-pVnZ+2d1f [11], and (most recently) cc-pV(n+d)Z[12] basis sets for second-row atoms exhibiting 'inner polarization' [11] (vide infra).

We consider here the following sequence of correlation consistent basis sets: A′VDZ+2d, A′VTZ+2d1f, A′VQZ+2d1f, and A′V5Z+2d1f, which we shall denote "small", "medium", "large", and "extra large" (for first-and second-row compounds, these basis sets are of spd, spdf, spdfg, and spdfgh quality, respectively). W1 theory, then, carries out all extrapolations using "small", "medium", and "large", while W2 theory employs "medium", "large", and "extra-large" basis sets.

The W1 and W2 protocols for obtaining the total atomization energy (TAE) of a given molecule involve the following steps:

1. Geometry optimization at the B3LYP/VTZ+1 level for W1, and at the CCSD(T)/VQZ+1 level for W2.

2. Extrapolation of the SCF component of TAE from the "small", "medium", and "large" basis sets (W1) or "medium", "large", and "extra-large" basis sets (W2), by means of either the geometric

extrapolation formula $E(n) = E_\infty + A/B^n$ (old-style) or the two-point formula $E(n) = E_\infty + A/n^5$ (new-style).

3. Extrapolation of the CCSD valence correlation component of TAE from the "medium" and "large" basis sets (W1) or from the "large" and "extra-large" basis sets (W2) employing the two-point formula $E(n) = E_\infty + A/B^\alpha$, where $\alpha = 3.22$ (W1) or 3 exactly (W2).

4. Extrapolation of the contribution to TAE of the connected triple excitations, (T), from the valence orbitals using the same formulae as for CCSD, but employing instead the "small" and "medium" basis sets (W1) or the "medium" and "large" basis sets (W2).

5. The contribution of inner-shell correlation is taken as the difference between the CCSD(T)/MTsmall TAE with and without constraining the inner-shell orbitals to be doubly occupied.

6. The scalar relativistic contribution is computed as the first-order Darwin and mass-velocity corrections from the ACPF/MTsmall wave function, including inner-shell correlation.

7. The contribution to TAE of spin-orbit splitting in the constituent atoms is trivially obtained from a tabulation, while for molecules in degenerate ground states, CISD/MTsmall spin-orbit splittings are computed (allowing correlation from the $2s$ and $2p$ orbitals in second-row atoms).

8. The zero-point vibrational energy (E_{ZPV}) is obtained from harmonic B3LYP/VTZ+1 frequencies scaled by 0.985 in the case of W1 theory. For W2 theory, anharmonic values of E_{ZPV} from quartic force fields at the CCSD(T)/VQZ+1 (or comparable) level are preferred; where this is not feasible, the same procedure as for W1 theory is followed as a "fallback solution".

We shall now proceed to explain in detail these steps and the rationale behind them.

2.1. Reference Geometry

Near the equilibrium geometry, dependence of the energy on geometric displacements is approximately quadratic. As a result, small errors in the reference geometry will insignificantly affect computed energies, but more substantial errors (say, several hundredths of an Å in covalent bond lengths) will compromise the reliability of a thermochemical calculation.

For W1 theory, we chose B3LYP [13, 14] density functional theory with the VTZ+1 basis set as the level of theory for the reference geometry, where the +1 suffix denotes the addition to second-row atoms of the highest-exponent *d* function from the V5Z basis set [10]. For first-row molecules, B3LYP/VTZ bond lengths are generally within 0.003 Å from experiment [15]; for second-row molecules, significant errors can be seen [10, 16] unless a tight *d* function is added to the basis set to account for inner polarization (see below).

For W2 theory, we opted for CCSD(T)/VQZ+1 as the level of theory for reference geometries. For geometries, the VQZ basis set is known to be close to the one-particle basis set limit [17, 18], while the addition of the inner polarization functions again takes care of inner polarization effects.

2.2. The SCF Component of TAE

For systems devoid of nondynamical correlation effects, this is the largest individual contribution to the molecular binding energy. Its basis set convergence is relatively rapid, yet our discussion will be disproportionately long because a number of the "dramatis personae" that reappear in the remainder of the story need to be introduced here.

For the SCF energy, we can – at least for small systems – obtain an exact answer by means of numerical SCF calculations. There is substantial empirical evidence that its convergence behavior is exponential. Jensen studied the SCF convergence behavior of the SCF energy in H_2 [19] and H_3^+ and N_2 [20] and found clear evidence of geometric convergence behavior in terms of both the maximum angular momentum in the basis set and the number of primitives within a given angular momentum.

Martin and Taylor [21] compared numerical SCF energies with extrapolations from calculated SCF/A'VQZ, SCF/A'V5Z, and SCF/A'V6Z energies using the formula

$$\mathrm{E(L)} = \mathrm{E}_\infty + \mathrm{A}/\mathrm{B}^\mathrm{L} \tag{2.1}$$

(which is equivalent to $\mathrm{E(L)} = \mathrm{E}_\infty + \mathrm{A}\exp(-\mathrm{BL})$ originally proposed by Feller [22]) and, for a number of number of molecules, found discrepancies of 10 $\mu\mathrm{E_h}$ or less between the numerical and extrapolated values.

Petersson et al. had earlier proposed [23] an alternative expression $\mathrm{E(n)} = \mathrm{E}_\infty + \sum_{\mathrm{l=n+1}}^{\infty} \mathrm{A}/(\mathrm{l}+1/2)^6$ in the context of the CBS methods developed in his group. The summation is carried out numerically in that paper, but in fact an elegant analytical approximation exists for summations of this type:

$$\sum_{m=L+1}^{\infty} A/(m+1/2)^n = \frac{A\psi^{(n-1)}(L+3/2)}{(n-1)!}, \tag{2.2}$$

where $\psi^{(n)}(x)$ represents the order n polygamma function [24] of x. Its asymptotic expansion has the leading terms

$$\begin{aligned}\psi^{(n)}(x) &= (-1)^{n-1}\left[\frac{(n-1)!}{x^n}+\frac{n!}{2x^{n+1}}+O(x^{-n-2})\right] \\ &= (-1)^{n-1}\frac{(n-1)!}{(x-1/2)^n}+O(x^{-n-2}).\end{aligned} \tag{2.3}$$

Hence

$$\begin{aligned}\frac{A\psi^{(n-1)}(L+3/2)}{(n-1)!} &= \frac{(-1)^{n-2}A(n-2)!}{(n-1)!(L+1)^{n-1}}+O(L^{-(n+1)}) \\ &\approx \frac{A}{(n-1)(L+1)^{n-1}}.\end{aligned} \tag{2.4}$$

This suggests the simple extrapolation formula $E(n) = E_\infty + A/n^5$, i.e. $E_\infty = E(n) + \frac{E(n)-E(n-1)}{(n/n-1)^5-1}$, where n is identified with the "n-tuple zetaness" of the Dunning correlation consistent VnZ basis sets. (For hydrogen and helium, n equals the maximum angular momentum plus one; for the main group elements it is equal to the maximum angular momentum). While an argumentation in favor of the Petersson-type formula can be built on the convergence behavior of triplet-coupled pairs, neither this formula nor the geometric one have a solid formal basis.

Fortunately, convergence on the SCF component of atomization energies is even more rapid than for the total energies; Martin and Taylor found for 14 first-row molecules [25] that differences between unextrapolated SCF/A'V5Z, geometrical extrapolations from SCF/A'V{T,Q,5}Z, and $A + B/L^5$ extrapolations from SCF/A'V{Q,5}Z results are on the order of 0.01 kcal/mol. For the method that we designated W2, which uses this basis set sequence, the choice of SCF extrapolation method is largely a non-issue. For the method that we designated W1, however, the geometric formula entails the use of results from the comparatively small A'VDZ basis set, which compromises the reliability of extrapolated SCF limits in systems with slow basis set convergence. In some cases (see Table 1 in Ref. 26), these can lead to errors of several kcal/mol. In addition, the two-point $A + B/L^5$ formula has the elegant property that it becomes immaterial whether the extrapolation is carried out on a reaction energy or on the individual absolute energies.

In the original W1/W2 paper [1], we opted for the geometric formula in view of the observed geometric convergence behavior. In a subsequent validation study [26] on a much wider variety of systems, we however found the two-point formula to be much more reliable, and we have adopted it henceforth.

Finally, an issue that arises with second-row systems should be addressed. It was first noted by Bauschlicher and Partridge [27] that the atomization energy of SO_2 is exceedingly sensitive to the presence of high-exponent d and f functions in the basis set. This phenomenon was ascribed to hypervalence; Martin and Uzan [10], however, found that the same phenomenon exists in systems that cannot be considered hypervalent by the wildest stretch of the imagination, like AlF. In addition, it was found [11, 16] that properties other than the energy are affected as well, with (e.g. in SO_2 [11] and SO_3 [16]) errors of up to 50 cm^{-1} in harmonic frequencies and hundredths of Å in bond lengths unless high-exponent d and f functions (termed "inner polarization functions" in Ref. 11 are added to the basis set.

We should note that inner polarization is strictly an SCF-level effect: while, for instance, switching from an A′VDZ to an A′VDZ+2d basis set affects the computed atomization energy of SO_3 by as much as 40 kcal/mol (!), almost all of this effect is seen in the SCF component of the TAE [28]. In fact, we have recently found [29] that the effect persists if the $(1s, 2s, 2p)$ orbitals on the second-row atom are all replaced by a pseudopotential. What is really getting "polarized" here is the inner part of the valence orbitals, which requires polarizations functions that are much "tighter" (higher-exponent) than those required for the outer part of the valence orbital. The fact that these inner polarization functions are in the same exponent range as the d and f functions required for correlation out of the $(2s, 2p)$ orbitals is merely coincidental; the "inner polarization" effect has nothing to do with correlation, let alone with inner-shell correlation.

After extensive numerical experimentation, we have decided [1] on the sequence of basis sets noted above: "small" A′VDZ+2d, "medium" A′VTZ+2d1f, "large" A′VQZ+2d1f, and "extra large" A′V5Z+2d1f.

As the present review was being finalized for publication, we received a preprint by Dunning et al. [12] where new cc-pV(n+d)Z basis sets are proposed for the second-row atoms. These basis sets do have just an added tight d function (hence the acronym) and no tight f functions, but the remaining d functions in the underlying cc-pVnZ basis set are in addition reoptimized. We are currently investigating their performance in W1 and W2-type schemes.

2.3. The CCSD Valence Correlation Component of TAE

The valence correlation component of TAE is the only one that can rival the SCF component in importance. As is well known by now (and is a logical consequence of the structure of the exact nonrelativistic Born-Oppenheimer Hamiltonian on one hand, and the use of a Hartree-Fock reference wavefunction on the other hand), molecular correlation energies tend to be dominated by double excitations and disconnected products thereof. Single excitation energies become important only in systems with appreciable nondynamical correlation. Nonetheless, since the number of single-excitation amplitudes is so small compared to the double-excitation amplitudes, there is no point in treating them separately.

For all intents and purposes then, we are concerned here with the CCSD (coupled cluster with all single and double substitutions [30]) correlation energy. Its convergence is excruciatingly slow: Schwartz [31] showed as early as 1963 that the increments of successive angular momenta l to the second-order correlation energy of helium-like atoms converge as

$$\Delta E(l) = A/(l+1/2)^4 + B/(l+1/2)^6 + \dots \quad (2.5)$$

His conclusions were generalized to other methods and general pair correlation energies by Hill [32] and by Kutzelnigg and Morgan [33].

This clearly spells a rather bleak picture of basis set convergence. Indeed, Martin [17] showed in 1994 that while convergence of σ bond energies appeared in sight at the CCSD(T)/spdfg level, this did not yet appear to be the case for π bond energies. This earlier study was extended in 1996 [34] to basis sets of spdfgh quality: somewhat depressingly, residual errors in the binding energies as high as 2 kcal/mol were still found for small systems.

However, rather than "knuckling under" to Eq.(2.5) at this stage, we might instead *exploit* it for an extrapolation formula. Martin [34] suggested a three-point extrapolation of the form $A + B/(n + 1/2)^C$ (where n is identified with the cardinal number of the cc-pVnZ basis set), and obtained dramatically improved computed total atomization energies. A slight further improvement was achieved if the SCF and valence correlation energies – which have fundamentally different convergence behaviors – are extrapolated separately using the respective appropriate formulae [25].

The denominator shift of 1/2 was chosen as a compromise between the situation for hydrogen and helium (where n = l + 1 for the cc-pVnZ basis set) and main-group elements (where n = l). As is immediately obvious upon series expansion, there is considerable coupling between the denominator shift and the exponent. As a result, the three-point extrapolation generally leads to exponents well in excess of three [34].

Halkier et al. [35] found the simple expression $E(L) = E_\infty + A/L^3$ [i.e. $E_\infty = E(L) + \frac{E(L)-E(L-1)}{(L/L-1)^3-1}$] to work at least equally well. In view of its simplicity and the fact that no results with the questionable A′VDZ basis set are required, we have adopted this simple formula for extrapolation of the CCSD valence correlation energy in W1 and W2 theories.

For the smaller basis sets used in W1 theory, the regime where the leading $E_\infty + A/L^3$ term dominates convergence behavior has not yet been reached, and using the formula in its unmodified form leads to overestimated (in absolute value) CCSD limits. One unelegant solution would be the use of three-term extrapolations like $E_\infty + A/L^3 + B/L^4$, but in light of the poor quality of the VDZ basis set this is a most unsatisfactory alternative. Another alternative is the use of a two-point extrapolation $E_\infty + A/L^\alpha$, in which α is a fixed empirical parameter. By minimizing the deviation from the W2 CCSD limit for the so-called W2-1 set of 28 molecules (vide infra), we determined $\alpha = 3.22$, which is the value used in W1 theory and its variants.

2.4. Connected Triple Excitations: the (T) Valence Correlation Component of TAE

It has been well known for some time (e.g. [36]) that the next component in importance is that of connected triple excitations. By far the most cost-effective way of estimating them has been the quasiperturbative approach known as CCSD(T) introduced by Raghavachari et al. [37], in which the fourth-order and fifth-order perturbation theory expressions for the most important terms are used with the converged CCSD amplitudes for the first-order wavefunction. This account for substantial fractions of the higher-order contributions; a very recent detailed analysis by Cremer and He [38] suggests that 87, 80, and 72 %, respectively, of the sixth-, seventh-, and eighth-order terms appearing in the much more expensive CCSDT-1a method are included implicitly in CCSD(T).

Nevertheless, the formidable n^3N^4 (with n the number of electrons and N the number of basis functions) cost scaling of the CCSD(T) method creates a substantial barrier to applications of methods that require A′V5Z+2d1f basis sets. However, two things should be kept in mind. First of all, the (T) component of TAE is a small fraction of the CCSD component, and hence a larger relative error can be tolerated. Secondly, evidence exists [39] that basis set convergence of the (T) contribution is substantially more rapid than that of the CCSD energy.

As a result, one may justifiably extrapolate the (T) contribution from smaller basis sets than its CCSD counterpart: in W1 theory, we extrapolate from the "small" and "medium" basis sets, and in W2 theory from the "medium" and "large" basis sets. This means that the most extensive basis sets in the calculations, namely "large" in W1 theory and "extra large" in W2 theory only require CCSD calculations, which are both much less expensive than CCSD(T) and much more amenable to direct algorithms such as those described in Refs. 40-41.

2.5. The Inner-Shell Correlation Component of TAE

Inner-shell correlation is a substantial part of the absolute correlation energy even for late first-row systems; for second-row systems, it in fact rivals the absolute valence correlation energy in importance. However, its relative contribution to molecular TAEs is fairly small: in benzene, for instance, it amounts to less than 0.7 % of the TAE. Even so, at 7 kcal/mol, its contribution is important by any reasonable thermochemical standard. By the same token, a 1 % relative error in a 7 kcal/mol contribution is tolerable even by benchmark thermochemistry standards, while the same relative error in a 300 kcal/mol contribution would be unacceptable even by the "chemical accuracy" standards.

In addition, for thermochemical purposes we are primarily interested in the core-valence correlation, since we can reasonably expect the core-core contributions to largely cancel between the molecule and its constituent atoms. (The partitioning between core-core correlation – involving excitations only from inner-shell orbitals – and core-valence correlation – involving simultaneous excitations from valence and inner-shell orbitals – was first proposed by Bauschlicher, Langhoff, and Taylor [42]).

For these reasons, we feel justified in treating the inner-shell correlation contribution to TAE as a separate contribution, rather than together with the valence correlation. There are substantial cost advantages to this: rather than having to carry out very elaborate all-electrons-correlated CCSD(T) calculations in basis sets near saturation for both valence and inner-shell correlation, we can limit these costly calculations to a basis set that is primarily saturated for inner-shell correlation.

Inner-shell correlation contributions for the W2-1 set were studied in some detail in the original W1/W2 paper, while subsequently, Martin, Sundermann, Fast, and Truhlar (MSFT) [43] studied inner-shell correlation contributions to TAE for 125 molecules spanning the first two rows of the periodic table. The following conclusions can be drawn from these two studies: (a) the use of the CCSD(T) electron correlation method is absolutely required for reliable contributions: the use of MP2 or CCSD can lead to underestimates in the order of 50 %; (b) the smallest basis set which gives acceptable agreement with near-basis set limit contributions is the MTsmall basis set, which is a completely decontracted cc-pVTZ basis set with (2d1f) additional high-exponent correlation functions; (c) the effect of including even higher excitations in the correlation treatment is insignificant.

A tentative explanation for the importance of connected triple excitations for the inner-shell contribution to TAE can be found in the need to account for simultaneously correlating a valence orbital and relaxing an inner-shell orbital, or conversely, requiring a double and a single excitation simultaneously.

In principle, one could contract at least the few innermost s primitives and reduce the basis set further. By leaving the basis set completely uncontracted, however, we can recycle the integrals and SCF wavefunction for the next step of the calculation.

Finally, it is generally advised *not* to correlate the very deep-lying ($1s$) orbitals on second-row elements, as the MTsmall basis set does not have angular correlation functions in the required exponent range, and in addition the orbitals concerned are in the same energy range as the ($2s, 2p$) orbitals in third-row main group elements, for which being able to take a [Ne] core out of the correlation problem does result in appreciable CPU time savings.

2.6. Scalar Relativistic Correction

The importance of scalar relativistic effects for compounds of transition metals and/or heavy main group elements is well established by now [44]. Somewhat surprisingly (at first sight), they may have non-trivial contributions to the TAE of first-row and second-row systems as well, in particular if several polar bonds to a group VI or VII element are involved. For instance, in BF_3, SO_3, and SiF_4, scalar relativistic effects reduce TAE by 0.7, 1.2, and 1.9 kcal/mol, respectively – quantities which clearly matter even if only "chemical accuracy" is sought. Likewise, in a benchmark study on the electron affinities of the first-and second-row atoms [45] – where we were able to reproduce the experimental values to within 0.001 eV on average – we saw that neglect of the scalar relativistic contributions increased mean deviation from experiment by more than an order of magnitude.

Perhaps the simplest and most cost-effective way of treating relativistic contributions in an all-electron framework is the first-order perturbation theory of the one-electron Darwin and mass-velocity operators [46, 47]. For variational wavefunctions, these contributions can be evaluated very efficiently as expectation values of one-electron operators.

It has been found repeatedly [1, 43, 45] that scalar relativistic contributions are overestimated by about 20 - 25 % in absolute value at the SCF level. Hence inclusion of electron correlation is essential: we found the ACPF method (which is both variational and approximately size extensive) to be an excellent compromise between quality and cost. It is reasonable to suppose that for a property that becomes more important as one approaches the nucleus, one wants maximum flexibility of the wavefunction near the nucleus as well as correlation of all electrons; thus we finally opted for ACPF/MTsmall as our approach of choice. Typically the cost of the scalar relativistic step is a fairly small fraction of that of the core correlation step, since only n^2N^4 scaling is involved in the ACPF calculations.

Bauschlicher [48] compared a number of approximate approaches for scalar relativistic effects to Douglas-Kroll quasirelativistic CCSD(T) calculations. He found that the ACPF/MTsmall level of theory faithfully reproduces his more rigorous calculations, while the use of non-size extensive approaches like CISD leads to serious errors. For third-row main group systems, studies by the same author [49] indicate that more rigorous approaches may be in order.

2.7. Spin-Orbit Coupling

The other relativistic effect entirely neglected so far is the spin-orbit coupling. For systems in nondegenerate states, the only first-order contribution to TAE comes from the fine structures in the corresponding atoms. Their effects can trivially be obtained from the observed electronic spectra, and hence the computational cost of this correction is fundamentally zero.

For systems in degenerate states, first-order corrections may need to be computed. In our work [26] we found that this significantly reduced the mean absolute error for the G2-1 and G2-2 test sets for ionization potentials and electron affinities, in no small part due to the preponderance of atoms and linear molecules in these sets. We found that CISD/MTsmall generally yields quite satisfactory spin-orbit corrections, but that it is advisable to correlate the $(2s, 2p)$-like electrons in the second-row elements. For the halogen atoms, convergence of these contributions with the level of theory was studied in some detail by Nicklass et al. [50]. These authors came to fundamentally the same conclusions.

2.8. The Zero-Point Vibrational Energy

It has been noted repeatedly (e.g. [51, 52, 53]) that one-half the sum of the harmonic frequencies, $\frac{1}{2}\sum_i \omega_i \, d_i$ (with d_i representing the degeneracy of mode i) generally leads to an overestimate of the E_{ZPV}, and that one-half the sum of the fundamentals, $\frac{1}{2}\sum_i \nu_i \, d_i$, generally leads to an underestimate. In fact, it is easily shown that the average of these two estimates is a fairly good approximation to the anharmonic E_{ZPV}.

For the sake of convenience, we shall restrict ourselves to the case of symmetric tops, asymmetric tops being a special case thereof with no degenerate modes. Including only up to first-order anharmonicities X_{ij}, and excluding the small constant E_0, the vibrational energy is given as

$$G(\mathbf{n}, \mathbf{l}) = \sum_i \omega_i \, (n_i + \frac{d_i}{2}) + \sum_{i \le j} X_{ij} \, (n_i + \frac{d_i}{2})(n_j + \frac{d_j}{2}) + S(\mathbf{l}) \, , \tag{2.6}$$

in which S is the splitting term involving the angular momenta l of the degenerate vibrations, and n_i represents the vibrational quantum number for mode i. It trivially follows that the zero-point energy E_{ZPV} is given by

$$E_{ZPV} = \sum_i \omega_i \, \frac{d_i}{2} + \sum_{i \le j} X_{ij} \, \frac{d_i \, d_j}{4} \, . \tag{2.7}$$

In addition we find that [introducing the shorthand $G(\mathbf{n}, \mathbf{l})^0 \equiv G(\mathbf{n}, \mathbf{l}) - G(0))$]

$$\begin{aligned} G(\mathbf{n}, \mathbf{l})^0 &= \sum_i \omega_i \, n_i + \sum_{i \le j} X_{ij} \, [(n_i + \frac{d_i}{2})(n_j + \frac{d_j}{2}) - \frac{d_i d_j}{4}] + S(\mathbf{l}) \\ &= \sum_i \omega_i \, n_i + \sum_{i \le j} X_{ij} \, [n_i n_j + n_i \frac{d_j}{2} + n_j \frac{d_i}{2}] + S(\mathbf{l}) \\ &= \sum_i \omega_i \, n_i + \sum_i X_{ii} \, n_i \, (n_i + d_i) \\ &\qquad + \frac{1}{2} \sum_{i \ne j} X_{ij} \, [n_i n_j + n_i \frac{d_j}{2} + n_j \frac{d_i}{2}] + S(\mathbf{l}) \, . \end{aligned} \tag{2.8}$$

Now assume only n_k is nonzero, then

$$
\begin{aligned}
G(n_k, l_k)^0 &= \omega_k\, n_k + X_{kk}\, n_k\, (n_k + d_k) \\
&\quad + \frac{1}{2} \sum_{i \neq k} (X_{ik} + X_{ki}) \frac{n_k d_i}{2} + S(l_k) \\
&= \omega_k\, n_k + X_{kk}\, n_k\, (n_k + d_k) + \sum_{i \neq k} X_{ik}\, n_k\, \frac{d_i}{2} + G_{kk}\, l_k^2 .
\end{aligned} \tag{2.9}
$$

It then follows that

$$
\begin{aligned}
\sum_k \nu_k \frac{d_k}{2} &= \sum_k \omega_k \frac{d_k}{2} + \sum_k X_{kk} \frac{d_k(1 + d_k)}{2} + \sum_k \sum_{i \neq k} X_{ik} \frac{d_i d_k}{4} \\
&\quad + \sum_k^{\text{degen.}} \frac{d_k}{2} G_{kk}\, l_k^2 \\
&= \sum_k \omega_k \frac{d_k}{2} + \sum_k X_{kk} \frac{d_k^2}{2} + \sum_k X_{kk} \frac{d_k}{2} + \sum_{k>i} X_{ik} \frac{d_i d_k}{2} \\
&\quad + \sum_k^{\text{degen.}} \frac{d_k}{2} G_{kk}\, l_k^2 \\
&= \sum_k \omega_k \frac{d_k}{2} + \sum_{k \geq i} X_{ik} \frac{d_i d_k}{2} + \sum_k X_{kk} \frac{d_k}{2} + \sum_k^{\text{degen.}} \frac{d_k}{2} G_{kk}\, l_k^2 .
\end{aligned} \tag{2.10}
$$

That is,

$$
\begin{aligned}
\sum_k (\nu_k + \omega_k) \frac{d_k}{4} &= \sum_k \omega_k \frac{d_k}{2} + \sum_{k \geq i} X_{ik} \frac{d_i d_k}{4} + \sum_k X_{kk} \frac{d_k}{4} \\
&\quad + \sum_k^{\text{degen.}} \frac{d_k}{4} G_{kk}\, l_k^2 \\
&= E_{ZPV} + \sum_k X_{kk} \frac{d_k}{4} + \sum_k^{\text{degen.}} \frac{d_k}{4} G_{kk}\, l_k^2 ,
\end{aligned} \tag{2.11}
$$

in which the G_{kk} are the diagonal l-coupling constants. The last term is generally negligible. If so desired, the term involving the diagonal anharmonicity constants can be estimated from anharmonicities in diatomic molecules.

The common practice of scaling computed vibrational frequencies for comparison with experimental fundamentals attempts at approximately addressing two issues: (a) the imperfections of the theoretical model for the harmonic frequency (which for CCSD(T), or even B3LYP, in sufficiently large basis sets is basically unnecessary); and (b) the anharmonic contribution to the fundamental. The above analysis suggests that a scaling factor that is intermediate between those used for reproducing harmonics and fundamentals would be the most appropriate for anharmonicities. In the original W1 paper [1], we considered the essentially exact anharmonic values of E_{ZPV} of the 28 W2-1 molecules (determined from experiment or large basis set CCSD(T) quartic force field calculations, e.g. [54] and the references therein) and found the appropriate scaling factor for B3LYP/VTZ+1 harmonic frequencies to be 0.985. The largest individual deviation between the scaled harmonic and exact anharmonic values of E_{ZPV} was only 0.3 kcal/mol (for PH_3).

Some of the above remarks are probably best illustrated by an example. For benzene, a B3LYP/TZ2P quartic force field was computed by Handy and coworkers [55]. From the published anharmonicity constants (specifically, the set deperturbed for Fermi resonances closer than 100 cm^{-1}), we obtain an anharmonic E_{ZPV} of 62.04 kcal/mol. For comparison, one-half the sum of the harmonics comes out 0.9 kcal/mol too high at 62.96 kcal/mol, and one-half the sum of the fundamentals comes out 1 kcal/mol too low at 60.98 kcal/mol. The average of both values, 61.97 kcal/mol, is in excellent agreement with the anharmonic value, while the W1 estimate accidentally agrees to within two decimal places with the B3LYP/TZ2P anharmonic value. From the best available computed harmonic frequencies [56] and the best available experimental fundamentals [55], we obtain E_{ZPV} = 62.01 kcal/mol or, after correction for the difference between this estimate and the true anharmonic E_{ZPV} at the B3LYP/TZ2P level, equal to 0.07 kcal/mol, we find E_{ZPV} = 62.08 kcal/mol as possibly the best estimate. (Note that HF/6-31G* harmonic frequencies scaled by 0.8929, as used in G2 and G3 theories, yields only 60.33 kcal/mol. In this accuracy range, one certainly cannot indulge in a 1.7 kcal/mol underestimate in the zero-point energy!)

In a recent benchmark study [57] on the CH_2=NH molecule, we explicitly computed a CCSD(T)/VTZ quartic force field at great expense (the low symmetry necessitated the computation of 2241 energy points in C_s symmetry and 460 additional points in C_1 symmetry). The resulting anharmonic E_{ZPV}, 24.69 kcal/mol, is only 0.10 kcal/mol above the scaled B3LYP/VTZ estimate, 24.59 kcal/mol. At least for fairly rigid molecules, it appears hard to justify the additional expense and effort for the anharmonic force field unless it were required anyway for other purposes.

If we use B3LYP/VTZ+1 harmonics scaled by 0.985 for the E_{ZPV} rather than the actual anharmonic values, mean absolute error at the W1 level deteriorates from 0.37 to 0.40 kcal/mol, which most users would regard as insignificant. At the W2 level, however, we see a somewhat more noticeable degradation from 0.23 to 0.30 kcal/mol – if kJ/mol accuracy is required, literally "every little bit counts". If one is primarily concerned with keeping the maximum absolute error down, rather than getting sub-kJ/mol accuracy for individual molecules, the use of B3LYP/VTZ+1 harmonic values of E_{ZPV} scaled by 0.985 is an acceptable "fallback solution". The same would appear to be true for thermochemical properties to which the E_{ZPV} contribution is smaller than for the TAE (e.g. ionization potentials, electron affinities, proton affinities, and the like).

3. PERFORMANCE OF W1 AND W2 THEORIES

A reliable assessment of the performance of a method in the kJ/mol accuracy range is, by its very nature, only possible where experimental data are themselves known to this accuracy.

3.1. Atomization Energies (the W2-1 Set)

In the original W1/W2 paper [1], we selected a set of 28 first-and second-row molecules (which we shall call the W2-1 set) containing at most three nonhydrogen atoms for which (a) the experimental total atomization energies $\sum D_0$ are available to the highest possible accuracy (preferably 0.1 kcal/mol); (b) no strong nondynamical correlation effects exist that would hinder the applicability of single-reference electron correlation methods; (c) near-exact anharmonic values of E_{ZPV} are available from either experimental anharmonicity constants or highly accurate ab initio anharmonic force fields.

Results using W1 and W2 theories are shown in Table 2.1. For W2 theory we find a mean absolute deviation (MAD) of 0.23 kcal/mol, which further drops to 0.18 kcal/mol when the NO, O_2, and F_2 molecules are deleted (all of which have mild nondynamical correlation in common). Our largest deviation is 0.70 kcal/mol. We can hence state that W2 meets our design goals.

Table 2.1 Comparison of W2 and W1 theories, and their variants for the evaluation of TAE_0 (kcal/mol) for the W2-1 test set.

Species	Experimental[a]		Deviation (experiment − theory)					
	TAE_0	± (uncert.)	W2[b]	W2[c]	W2h[d]	W1	W1h[d]	W1c
H_2	103.27	0.00	-0.05	-0.04		-0.07		-0.07
N_2	225.06	0.04	0.36	0.45		0.53		0.54
O_2	117.97	0.04	0.64	0.68		0.41		0.18
F_2	36.94	0.10	0.60	0.78		0.70		0.52
HF	135.33	0.17	0.02	-0.07		-0.47		-0.41
CH	79.90	0.23	-0.08	-0.15	-0.14	-0.17	-0.11	-0.37
CO	256.16	0.12	0.12	0.12	0.14	-0.08	-0.06	-0.41
NO	149.82	0.03	0.47	0.54		0.56		0.33
CS	169.41	0.23	0.30	0.31	0.32	0.77	0.95	0.46
SO	123.58	0.04	-0.02	-0.04		0.52		0.57
HCl	102.24	0.02	-0.04	-0.14		-0.15		-0.17
ClF	60.36	0.01	0.09	0.08		0.15		0.03
Cl_2	57.18	0.00	-0.20	-0.24		0.60		0.50
HNO	196.85	0.06	0.38	0.37		0.20		-0.03
CO_2	381.91	0.06	0.14	0.13	0.10	-0.37	-0.34	-0.37
H_2O	219.35	0.12	-0.04	-0.14		-0.55		-0.58
H_2S	173.15	0.12	-0.37	-0.49		-0.47		-0.51
HOCl	156.61	0.12	-0.16	-0.24		-0.18		-0.40
OCS	328.53	0.48	-0.19	-0.21	-0.21	-0.01	0.11	0.10
ClCN	279.20	0.48	0.41	0.52	0.78	0.78	0.91	0.82
SO_2	253.92	0.08	-0.31	-0.33		0.63		0.81
CH_3	289.00	0.10	-0.21	-0.32	-0.38	-0.53	-0.51	-0.39
NH_3	276.73	0.13	0.13	-0.03		-0.28		-0.17
PH_3	227.13	0.41	-0.01	0.28		0.23		0.05
C_2H_2	388.90	0.24	0.42	0.64	0.53	0.26	0.51	0.29
CH_2O	357.25	0.12	-0.27	-0.40	-0.35	-0.59	-0.56	-0.76
CH_4	392.51	0.14	-0.11	-0.13	-0.19	-0.35	-0.47	-0.34
C_2H_4	531.91	0.17	-0.19	-0.31	-0.32	-0.63	-0.41	-0.72
Mean Absolute Deviation			0.23	0.29	0.30	0.40	0.41	0.39
Max. Absolute Deviation			0.64	0.78	0.78	0.78	0.95	0.82

[a] See [1] for experimental references.

[b] Values of E_{ZPV} derived from anharmonic vibrational frequencies. See Ref. 1 for details.

[c] Values of E_{ZPV} derived from B3LYP/VTZ+1 harmonic vibrational frequencies scaled by 0.985. Same remark applies to W2h, W1, W1h and W1c data given.

[d] For systems where W2h and W1h are equivalent to W2 and W1, respectively, entries have been left blank.

For W1 theory, MAD is increased to 0.37 kcal/mol (old SCF extrapolation) or 0.40 kcal/mol (new SCF extrapolation), with the maximum error being 0.78 kcal/mol. This should be compared with MAD of 1.25 kcal/mol for G2 theory, 0.89 kcal/mol for G3 theory, 0.88 kcal/mol for CBS-Q, and 0.61 kcal/mol for CBS-QB3, and the much higher maximum errors of these methods of 4.90 kcal/mol (SO_2), 3.80 kcal/mol (SO_2), 3.10 kcal/mol (OCS), and 1.90 kcal/mol (C_2H_2), respectively. While we would prefer to use W2 theory for no-nonsense benchmarking if at all possible, W1 theory still seems to offer great advantages over the other techniques.

3.2. Electron Affinities (the G2/97 Set)

Some representative results can be found in Table 2.2. For the G2-1 set of electron affinities, W1 theory has a mean absolute error of 0.016 eV [26]. Not unexpectedly – given the slow basis set convergence of electron affinities – the extra effort invested in W2 theory pays off with a further reduction of the mean absolute error to 0.012 eV. Accuracy appears to be limited principally by imperfections in the CCSD(T) method: for the atoms B–F and Al–Cl, using even larger basis sets we achieve 0.009 eV at the CCSD(T) level, which decreases to 0.001 eV if approximate full CI energies are used.

Normally W1 theory does not involve diffuse functions on H, Li, Na, Be, and Mg; not surprisingly, this leads to very poor electron affinities for Li and Na. Upon switching to W1aug (i.e. using augmented basis sets on all elements), perfect agreement with experiment is obtained. Within the G2-2 set, substantial discrepancies between W1 theory and experiment are found for O_3 and CH_2NC, both of which are systems that have pronounced multireference character. (The same remark applies to a lesser extent to FO.) Scalar relativistic effects almost invariably decrease the electron affinity. Neglect of spin-orbit splitting leads to significant deterioration in MAD.

3.3. Ionization Potentials (the G2/97 Set)

Some representative results can again be found in Table 2.2. At the W1 level, the G2-1 ionization potentials are reproduced with a MAD of only 0.013 eV [26]. No further improvement is seen at the W2 level for this property. Note that if the B3LYP/VTZ geometry for CH_4^+ is employed, a serious error is seen for IP(CH_4) which disappears when a CCSD(T)/VTZ reference geometry is used instead. (Only BH & HLYP

Table 2.2 Comparison of W2 and W1 theories, and their variants for the evaluation of electron affinity and ionization potential (eV) for selected species from G2-1 test set.

Species	Experimental[a]		Deviation (experiment − theory)			
	Value	± (uncert.)	W2	W2h	W1	W1h
			Electron Affinities			
C	1.2629	0.0003	0.007	0.041	0.011	0.210
Si	1.38946	0.00006	0.010	0.081	0.011	0.060
CH	1.238	0.0078	0.029	0.060	0.032	0.248
CH_2	0.652	0.006	0.002	0.042	0.011	0.236
CH_3	0.08	0.03	0.034	0.088	0.051	0.284
SiH	1.2771	0.0087	0.031	0.094	0.034	0.084
SiH_2	1.123	0.022	0.039	0.088	0.043	0.087
SiH_3	1.406	0.014	0.011	0.033	0.019	0.044
CN	3.862	0.005	-0.026	-0.036	-0.031	-0.023
			Ionization Potentials			
B	8.29802	0.00002	0.007	0.009	0.019	0.020
C	11.2603	0.0001	0.010	-0.002	0.012	0.012
Al	5.986	0.001	0.023	0.022	0.024	0.025
Si	8.15166	0.00003	0.018	-0.004	0.021	0.022
CH_4 (b)	12.61	0.01	-0.033	-0.035	-0.032	-0.035
SiH_4	11	0.02	0.006	0.006	-0.005	-0.005
C_2H_2	11.403	0.0003	-0.004	-0.004	-0.001	0.005
C_2H_4	10.5138	0.0006	-0.001	0.001	-0.005	0.000
CO	14.0142	0.0003	-0.014	-0.013	-0.009	-0.008
CS	11.33	0.01	-0.017	-0.018	-0.017	-0.016

[a] See Ref. 26 for experimental references.

[b] CCSD(T)/VTZ geometry. B3LYP/VTZ optimization erroneously yields D_{2d} structure for cation rather than correct C_{2v} symmetry. See Ref. 26 for details.

[58] and mPW1K [59] correctly predict a C_{2v} structure for CH_4^+; other exchange-correlation functionals wrongly lead to a D_2 structure).

Inner-shell correlation contributions are found to be somewhat more important for ionization potentials than for electron affinities, which is understandable in terms of the creation of a valence 'hole' by ionization into which inner-shell electrons can be excited. Again, inclusion of spin-orbit splitting is worthwhile.

3.4. Heats of Formation (the G2/97 Set)

A detailed discussion and a table can be found in Ref. 26. First of all, we note that the mean uncertainty for the experimental values in the G2-1 set is itself 0.6 kcal/mol. MAD values for W1 and W2 theory stand at 0.6 and 0.5 kcal/mol, respectively, suggesting that these theoretical methods have a reliability comparable to the experimental data themselves.

For a subset of 27 G2-2 molecules with fairly small experimental uncertainties, W1 theory had MAD of 0.7 kcal/mol, compared to the average experimental uncertainty of 0.4 kcal/mol. Some systems exhibit deviations from experiment in excess of 1 kcal/mol: in the cases of BF_3 and CF_4, very slow basis set convergence is responsible, and W2 calculations in fact remove nearly all remaining disagreement with experiment for the latter system. (The best available value for BF_3 is itself a theoretical one, so a comparison would involve circular reasoning.) Other molecules (NO_2 and ClNO) suffer from severe multireference effects.

3.5. Proton Affinities

For proton affinities, W1 theory can basically be considered converged [26]. The W2 computed values are barely different from their W1 counterparts, and the latter's MAD of 0.43 kcal/mol is well below the about 1 kcal/mol uncertainty in the experimental values. W1 theory would appear to be the tool of choice for the generation of benchmark proton affinity data for calibration of more approximate approaches.

4. VARIANTS AND SIMPLIFICATIONS

4.1. W1′ Theory

It was noted that the original W1 theory (old-style SCF extrapolation) performed considerably more poorly for second-row than for first-row species. This was ascribed to the lack of balance in the basis sets for second-row atoms used in the SCF and valence correlation steps of W1; in particular, the A′VTZ+2d1f basis set contains as many "tight" d and f functions as regular ones, which would appear to be a bit top-heavy.

It was proposed to replace the A′VTZ+2d1f basis set by A′VTZ+2d, a conclusion borne out by calculations on the SO_3 molecule [28], which suffers from extreme inner polarization effects and as such provides a good "proving ground".

Compared to its prototype, the modification (the so-called W1′ theory) did appear to yield improved results for second-row molecules. However, in the W1/W2 validation study [26] we found this to be an artifact of the exaggerated sensitivity of the (old-style) 3-point geometric SCF extrapolation. Use of the new-style $E_\infty + A/L^5$ extrapolation largely eliminates both the problem and the difference between W1 and W1′ theory.

4.2. W1h and W2h Theories

While the need for diffuse-function augmented basis sets for highly electronegative elements is well established (e.g. [34]), it could be argued that they are not really required on group III and IV elements. For organic-type molecules in particular, this would result in significant savings.

We define here W1h and W2h theories, respectively, as the modifications of W1 theory for which AVnZ basis sets are only used on elements of groups V, VI, VII, and VIII, but regular VnZ basis sets on groups I, II, III, and IV. (The "h" stands for "heteroatom", as we originally investigated this for organic molecules.) For the purpose of the present paper, we have repeated the validation calculations described in the previous section for W1h and W2h theories. (For about half of the systems, W1 and W1h are trivially equivalent.) Some representative results can be found in Table 2.1 for atomization energies/heats of formation, and in Table 2.2 for ionization potentials and electron affinities.

For the heats of formation in the G2-1 set, the largest difference between W1 and W1h theory is 0.3 kcal/mol for Si_2; the average difference is less than 0.1 kcal/mol. For some of the systems in the G2-2 set, however, differences are more pronounced, e.g. 0.6 kcal/mol for CF_4 and 0.8 kcal/mol for benzene. (Note that the benzene calculation reported as an example application in the original W1 paper [1] is in fact a W1h calculation: the remaining small difference between that reference and the present work is due to the different SCF extrapolations used.) For the G2-1 heats of formation, W2h and W2 are essentially indistinguishable in quality, as could reasonably be expected.

For the G2-1 ionization potentials, the largest differences are 0.005 and 0.006 eV, respectively, for ethylene and acetylene. Differences in the G2-2 set are likewise small, although Si_2H_2 (0.009 eV) and CH_3OF

(0.024 eV) stand out. Clearly W1h is of a quality comparable to W1 for ionization potentials, and we recommend it as a moderately inexpensive high-accuracy method for this property. (As noted before, W2 does not represent an improvement over W1 for ionization potentials, and the same goes for W2h theory.)

For electron affinities, the differences between W1h and W1 are very pronounced, and become (as expected) particularly large (e.g. 0.284 eV in CH_3) for species where none of the atoms carry diffuse functions in W1h theory. The differences between W2 and W2h theory are still quite sizable, and in fact agreement with experiment for W2h is inferior to that for the less expensive W1 method. In summary, we do not recommend W1h or W2h for electron affinities.

4.3. A Bond-Equivalent Model for Inner-Shell Correlation

In a pilot W1h calculation on benzene [1], it was found that 85 % of the CPU time was spent on the inner-shell correlation step. Given that this contribution is about 0.5 % of the TAE of benzene, the CPU time proportion appears to be lopsided to say the least. On the other hand, a contribution of 7 kcal/mol clearly cannot be neglected by any reasonable standard. However, inner-shell correlation is by its very nature a much more local phenomenon than valence correlation, and a relative error of a few percent in such a small contribution is more tolerable than a corresponding error in the major contributions, Martin, Sundermann, Fast and Truhlar (MSFT) [43] investigated the applicability of a bond equivalent model.

We started by generating a data base of inner-shell correlation contributions for some 130 molecules that cover the first two rows of the periodic table. In order to reduce the number of parameters in the model to be fitted, we introduced a Mulliken-type approximation for the parameters $D_{AB} \approx (D_A+D_B)/2$. Furthermore we did retain different parameters for single and multiple bonds, but assumed $D_{A\equiv B} \approx (3/2)D_{A=B}$.

The model (which requires essentially no CPU time) was found to work very satisfactorily; its performance for the W2-1 set can be seen in Table 2.3. Somewhat to our surprise, we found that the same model performs reasonably well when applied to the scalar relativistic contributions, albeit with larger individual deviations.

It was recently suggested by Nicklass and Peterson [60] that the use of core polarization potentials (CPPs) [61] could be an inexpensive and effective way to account for the effects of inner shell correlation. The great potential advantage of this indeed rather inexpensive method over the MSFT bond-equivalent model is that it does not depend on

Table 2.3 Comparison of core correlation contributions to TAE_0 (kcal/mol) for the W2-1 test set.

Species	CCSD(T)/ very large[a]	CCSD(T)/ MTsmall	MSFT model	CPP n = 1[b]	CPP n = 2[b]
H_2	0.00	0.00	0.00		
N_2	0.75	0.82	0.80	0.74	1.08
O_2	0.24	0.24	0.50	0.28	0.43
F_2	-0.09	-0.08	0.18	0.05	0.06
HF	0.18	0.18	0.09	0.10	0.19
CH	0.14	0.14	0.30	0.29	0.48
CO	0.94	0.90	1.26	0.76	1.12
NO	0.40	0.41	0.51	0.46	0.69
CS	0.75	0.66	1.08		
SO	0.46	0.42	0.38		
HCl	0.20	0.15	0.15		
ClF	0.08	0.09	0.23		
Cl_2	0.19	0.18	0.29		
HNO	0.40	0.41	0.68	0.41	0.69
CO_2	1.64	1.67	1.68	1.12	1.88
H_2O	0.37	0.37	0.36	0.20	0.39
H_2S	0.34	0.25	0.24		
HOCl	0.31	0.29	0.50		
OCS	1.68	1.58	1.49		
ClCN	1.76	1.71	1.73		
SO_2	0.67	0.78	0.68		
CH_3	1.04	1.04	0.89	0.37	0.84
NH_3	0.62	0.64	0.49	0.29	0.62
PH_3	0.30	0.22	0.35		
C_2H_2	2.44	2.34	2.38	1.17	2.17
CH_2O	1.25	1.26	1.44	0.65	1.24
CH_4	1.21	1.21	1.19	0.48	1.01
C_2H_4	2.36	2.27	2.38	1.02	2.02
Mean Absolute Deviation		0.04	0.12	0.39	0.19
Max. Absolute Deviation		0.11	0.33	1.34	0.34
C_6H_6		7.09	7.13		6.30

[a] See Ref. 1 for details.

[b] See Ref. 60 for details.

any explicit connectivity information. The different approximate treatments of inner-shell correlation are compared with large-scale CCSD(T) results for the W2-1 set in Table 2.3. As seen there, while the CPP approach is indeed quite promising (clearly superior to MP2 calculations, for instance), it clearly requires further refinement. The MSFT bond-equivalent model in fact outperforms all other approximate methods, with a computational cost that is essentially nil.

4.4. Reduced-Cost Approaches to the Scalar Relativistic Correction

The fact that the additivity model for the scalar relativistic correction worked *at all* is a pleasant surprise: yet alternatives clearly merit exploration. As noted above, the SCF-level scalar relativistic contributions of Kedziora et al. [62] are systematically overestimated. One possibility which suggests itself then would be applying a scaling factor to the SCF values: we have considered this approach for the set of 120 molecules for which ACPF/MTsmall data were generated by MSFT for the purposes of parameterizing their empirical model. However, rather than following the more elaborate approach of Kedziora et al., we simply evaluated the first-order Darwin and mass velocity corrections by perturbation theory. We considered variation of the basis set, and found not surprisingly that typical contracted VnZ basis sets are insufficiently flexible in the core region. We found VTZuc+1 (where VTZuc stands for an uncontracted cc-pVTZ basis set) to be the best compromise between cost and quality.

The best scale factor in the least-squares sense is 0.788; while the mean absolute error of 0.04 kcal/mol is more than acceptable, the maximum absolute error of 0.20 kcal/mol (for SO_2) is somewhat disappointing. Representative results (for the W2-1 set) can be found in Table 2.4.

This error can be considerably reduced, at very little cost, by employing B3LYP density functional theory instead of SCF. The scale factor, 0.896, is much closer to unity, and both mean and maximum absolute errors are cut in half compared to the scaled SCF level corrections. (The largest errors in the 120-molecule data set are 0.10 kcal/mol for P_2 and 0.09 kcal/mol for BeO.) It could in fact be argued that the remaining discrepancy between the scaled B3LYP/cc-pVTZuc+1 values is on the same order of magnitude as the uncertainty in the ACPF/MTsmall values themselves.

Table 2.4 Comparison of scalar relativistic effect contributions to TAE_0 (kcal/mol) for the W2-1 test set.

Species	ACPF/ MTsmall	MSFT model	B3LYP/ VTZuc+1 scaled 0.896	SCF/ VTZuc+1 scaled 0.788
H_2	0.00	0.00	0.00	0.00
N_2	-0.11	-0.14	-0.15	-0.16
O_2	-0.15	-0.30	-0.18	-0.22
F_2	0.03	-0.37	-0.04	-0.09
HF	-0.20	-0.19	-0.18	-0.20
CH	-0.03	-0.05	-0.04	-0.04
CO	-0.14	-0.33	-0.17	-0.19
NO	-0.16	-0.20	-0.20	-0.22
CS	-0.15	-0.29	-0.21	-0.25
SO	-0.31	-0.27	-0.34	-0.40
HCl	-0.26	-0.17	-0.25	-0.26
ClF	-0.12	-0.35	-0.16	-0.23
Cl_2	-0.15	-0.34	-0.19	-0.26
HNO	-0.24	-0.28	-0.27	-0.29
CO_2	-0.45	-0.44	-0.48	-0.50
H_2O	-0.26	-0.26	-0.25	-0.26
H_2S	-0.41	-0.43	-0.39	-0.40
HOCl	-0.28	-0.43	-0.31	-0.37
OCS	-0.53	-0.41	-0.57	-0.57
ClCN	-0.43	-0.40	-0.47	-0.47
SO_2	-0.71	-0.61	-0.79	-0.90
CH_3	-0.17	-0.14	-0.17	-0.16
NH_3	-0.25	-0.24	-0.25	-0.24
PH_3	-0.46	-0.60	-0.45	-0.46
C_2H_2	-0.27	-0.31	-0.28	-0.26
CH_2O	-0.32	-0.32	-0.33	-0.34
CH_4	-0.19	-0.19	-0.19	-0.18
C_2H_4	-0.33	-0.34	-0.33	-0.31
Mean Absolute Deviation		0.08	0.03	0.05
Max. Absolute Deviation		0.40	0.08	0.20

4.5. W1c Theory

Here we propose a new reduced-cost variant of W1 theory which we shall denote W1c (for "cheap"), with W1ch theory being derived analogously from W1h theory. Specifically, the core correlation and scalar relativistic steps are replaced by the approximations outlined in the previous two sections, i.e. the MSFT bond additivity model for inner-shell correlation and scaled B3LYP/cc-pVTZuc+1 Darwin and mass-velocity corrections. Representative results (for the W2-1 set) can be seen in Table 2.1; complete data for the molecules in the G2-1 and G2-2 sets are available through the World Wide Web as supplementary material [63] to the present paper.

As seen in Table 2.1, W1c is an acceptable "fallback solution" for systems for which W1 calculations are not feasible because of the number of inner-shell orbitals; for heats of formation and certainly for ionization potentials, W1ch offers a significant further cost reduction over W1h at a negligible loss in accuracy.

4.6. Detecting Problems

While CCSD and especially CCSD(T) are known [36] to be less sensitive to nondynamical correlation effects than low-order perturbation theoretical methods, some sensitivity remains, and deterioration of W1 and W2 results is to be expected for systems that exhibit severe nondynamical correlation character. A number of indicators exist for this, such as the $\mathcal{T}_1$ diagnostic of Lee and Taylor [64], the size of the largest amplitudes in the converged CCSD wavefunction, and natural orbital occupations of the frontier orbitals.

One pragmatic criterion which we have found to be very useful is the percentage of the TAE that gets recovered at the SCF level. For systems that are wholly dominated by dynamical correlation, like CH_4 and H_2, this proportion exceeds 80 %, while it drops to 50 % for the N_2 molecule, O_2 is only barely bound at the SCF level, and F_2 is even metastable. In the W1/W2 validation paper [26], we invariably found that large deviations from what appeared to be reliable experimental data tend to be associated with strong nondynamical correlation, and a small SCF component of TAE (e.g. 27 % for NO_2, 32 % for F_2O, and 15 % for ClO).

Would the use of full CCSDT [65] energies, instead of their quasi-perturbative-triples CCSD(T) counterparts, solve the problem? Our experience has taught us that this generally leads to a *deterioration* of the results; it has been shown (e.g. [66]) that the excellent performance of CCSD(T) for binding energies is at least in part due to error compensation between partial neglect of higher-order T_3 effects and complete neglect of T_4 effects. Unfortunately, explicit treatment of T_4 (connected quadruple excitations) is at present not feasible for practical-sized systems.

For some very small systems (e.g. Be_2 [67] and OH/OH^- [68]), we have considered what one might term W1CAS and W2CAS, in which the CCSD(T) calculations were replaced by full valence (or larger) CAS-ACPF calculations. The SCF extrapolation was then applied to the CASSCF (i.e. Hartree-Fock plus static correlation) energy, and the CCSD/CCSD(T) extrapolation to the dynamical correlation energy only. Aside from limited applicability due to the explosive increase in the number of reference configurations with the number of atoms, the formal objection of course applies that any separation between "internal" and "external" orbital spaces is to a large extent arbitrary.

Common sense also suggests that the larger the "gap" being bridged by the extrapolation from the actual computed number with the largest basis set to the hypothetical basis set limit, the larger the uncertainty in the latter will be. (See the example of benzene in section 5.3.)

Finally, the GIGO ("garbage in, garbage out") theorem applies here as well as in any other matter. For instance, if a B3LYP/cc-pVTZ+1 reference geometry is used for a system where the B3LYP geometry is known to be qualitatively wrong (such as CH_4^+), the computed W1 energetics will not be very reliable either.

5. EXAMPLE APPLICATIONS

5.1. Heats of Vaporization of Boron and Silicon

First-principle computation of gas-phase molecular heats of formation by definition requires the gas-phase heats of formation of the elements:

$$\begin{aligned}\Delta H^\circ_{f,T}(X_kY_l\cdots) \quad &- \quad k\,\Delta H^\circ_{f,T}(X) - l\,\Delta H^\circ_{f,T}(Y) - \cdots \\ = E_T(X_kY_l\cdots) \quad &+ \quad RT\,(1-k-l-\cdots) - k\,E_T(X) - l\,E_T(Y) - \cdots\,.\end{aligned} \tag{5.1}$$

Somewhat disappointingly, the values of $\Delta H^\circ_f[A(g)]$ of some first- and second-row elements A (notably boron and silicon) are not precisely known because of a variety of experimental difficulties. However, well-established precise heats of formation of BF_3 [69] and SiF_4 [70] are available that do not involve the heats of vaporization of boron and silicon in their determination. Thus, if accurate computed TAE_0 values of BF_3 and SiF_4 were available, then, in combination with the established value [71] of $D_0(F_2)$, the quantities sought for could be derived from a thermochemical cycle. These were obtained by means of W2 theory for BF_3 [72] and for SiF_4 [73]. The final recommended values are $\Delta H^\circ_{f,0}[B(g)] = 135.1 \pm 0.75$ kcal/mol and $\Delta H^\circ_{f,0}[Si(g)] = 107.15 \pm 0.38$ kcal/mol. The boron value is about 2 kcal/mol higher than the CODATA recommended value and in between a recent evaluation by Hildenbrand [74] and a 1977 measurement by Storms and Mueller [75]. The silicon value is slightly higher than the CODATA recommended value, and with a much smaller uncertainty. We note in passing that one of the first arguments for revision of $\Delta H^\circ_{f,0}[B(g)]$ and $\Delta H^\circ_{f,0}[Si(g)]$ was given in [76] on computational (CBS-Q) grounds.

5.2. Validating DFT Methods for Transition States: the Walden Inversion

It is well known (e.g. [77, 78]) that the prediction of reaction barrier heights is one of the main "Achilles' heels" of density functional theory. For instance [79], for the prototype S_N2 reaction,

$$X^- + CH_3Y \rightarrow CH_3X + Y^- , \tag{5.2}$$

B3LYP predicts a negative overall barrier if X = Y = Cl (i.e. a barrier between the entry and exit ion-molecule complexes that lies below the entrance channel). Adamo and Barone [79] demonstrated that their new mPW1PW91 (modified Perdew-Wang) functional at least yields the correct sign for this problem.

In Ref. 80 we carried out a W1 and W2 investigation for all six cases with $X,Y \in \{F, Cl, Br\}$, in order to assess the performance of a number of DFT exchange-correlation functionals. W2 is in excellent agreement with experiment where reliable experimental data are available; in some other cases, the W1 calculations either suggest revisions or provide the only reliable data available (see Ref. 80 for details).

Of the different exchange-correlation functionals considered, the new mPW1K [59] functional of Truhlar and coworkers appears to yield the best performance among "hybrid" functionals (i.e. those including a fraction of exact exchange), followed by BH&HLYP (a half-and-half mixture [58] of Hartree-Fock and Becke 1988 exchange [81] with Lee-Yang-Parr correlation). Among "pure DFT" functionals, the best performance is delivered by HCTH-120 [82] (the 120-molecule reparameterization of the Hamprecht-Cohen-Tozer-Handy functional). (We note in passing that this latter functional was parameterized entirely against ab initio data.) The G2 data of Pross et al. [83], despite some quantitative discrepancies, is qualitatively in perfect agreement with W1 theory.

We also note that in one case (F, Br) it was impossible to obtain all required stationary points at the B3LYP level, since the $F\cdots CH_3Br$ minimum does not show up at all at this level. Only mPW1K and BH&HLYP find this stationary point, as does CCSD(T).

5.3. Benzene as a "Stress Test" of the Method

As an illustrative example of "stress-testing" W1 and W2 theory, we shall consider the benzene molecule. The most accurate calculation we were able to carry out is at the W2h level: the rate-determining step was the direct CCSD/cc-pV5Z calculation (30 electrons correlated, 876 basis functions, carried out in the D_{2h} subgroup of D_{6h}) which took nearly two weeks on an Alpha EV67/667 MHz CPU. Relevant results are collected in Table 2.5.

At first sight, the disagreement between the computed W2h value of $\Delta H^\circ_{f,0K} = 23.0$ kcal/mol and the experimental value of 24.0±0.2 kcal/mol seems disheartening. (Note that it "errs" on the other side as the most recent previous benchmark calculation [53], 24.7±0.3 kcal/mol, using similar-sized basis sets as W1 theory.) However, the comparison with experiment is not entirely "fair" since it neglects the experimental uncertainties in the atomic heats of formation required to convert an atomization energy into a heat of formation (or vice versa). Combining these with the experimental $\Delta H^\circ_{f,0K}$ leads to an experimentally derived $TAE_0 = 1305.7 \pm 0.7$ kcal/mol, where the uncertainty is dominated by six times that in the heat of vaporization of graphite. In other words, our calculated $TAE_0 = 1306.7$ kcal/mol is only 0.3 kcal/mol removed from the upper end of the experimental uncertainty interval. (After all, an error of 0.02 % seems to be a bit much to ask for.)

Secondly, let us consider the "gaps" bridged by the extrapolations. For the SCF component, that gap is a very reasonable 0.3 kcal/mol (0.03 %), but for the CCSD valence correlation component this rises to 5 kcal/mol (1.7 %) while for the connected triple excitations contribution it amounts to 1 kcal/mol (3.7 % – note however that a smaller basis set is being used than for CCSD). It is clear that the extrapolations are indispensable to obtain even a *useful* result, let alone an accurate one, even with such large basis sets.

Inner-shell correlation, at 7 kcal/mol, is of quite nontrivial importance, but even scalar relativistic effects (at 1 kcal/mol) cannot be ig-

Table 2.5 Individual components (kcal/mol) in W1h, W1, and W2h total atomization energy *cum* heat of formation of benzene.[a]

Reference geometry	B3LYP/cc-pVTZ				CCSD(T)/cc-pVQZ	
	W1h		W1		W2h	
SCF	VDZ	1024.19	A'VDZ	1024.59	VTZ	1042.16
	VTZ	1042.10	A'VTZ	1042.62	VQZ	1044.62
	VQZ	1044.56	A'VQZ	1044.84	V5Z	1045.30
old-style	V∞Z	1044.95	V∞Z	1045.15	V∞Z	1045.56
new-style	V∞Z	1045.33	V∞Z	1045.53	V∞Z	1045.63
CCSD	VDZ	225.94	A'VDZ	226.11	VTZ	265.49
	VTZ	265.55	A'VTZ	268.44	VQZ	280.91
	VQZ	280.97	A'VQZ	282.39	V5Z	285.72
	V∞Z	291.08	V∞Z	291.53	V∞Z	290.77
(T)	VDZ	18.72	A'VDZ	19.64	VTZ	24.41
	VTZ	24.42	A'VTZ	24.78	VQZ	25.74
	V∞Z	26.55	V∞Z	26.69	V∞Z	26.71
Inner-shell correlation		7.09		7.08		7.10
Darwin and mass-velocity		-0.99		-0.99		-0.99
Spin-orbit coupling		-0.51		-0.51		-0.51
TAE_e		1368.54		1369.33		1368.71
E_{ZPV}		62.04		62.04		62.04
TAE_0		1306.49		1307.29		1306.67
$\Delta H^\circ_{f,0K}[C_6H_6(g)]$		23.18		22.39		23.01
$\Delta[H_{298.15} - H_0]$		-4.24		-4.24		-4.24
$\Delta H^\circ_{f,298.15K}[C_6H_6(g)]$		18.95		18.15		18.78

[a] Lower level TAE_0: 1301.9 (G2), 1305.2 (G3), 1303.7 (CBS-QB3), and 1304.3 (CBS-Q) kcal/mol. Experiment: $\Delta H^\circ_{f,0K}[C_6H_6(g)] = 24.0\pm0.2$ kcal/mol. [J. B. Pedley, *Thermodynamic Data and Structures of Organic Compounds* (Thermodynamics Research Center College Station, TX, 1994); Vol. 1.] This standard enthalpy of formation produces $TAE_0 = 1305.7\pm0.7$ kcal/mol, where the uncertainty equals $\sqrt{0.2^2 + (6 \times 0.11)^2}$; 0.11 kcal/mol being the uncertainty in the CODATA $\Delta H^\circ_{f,0}$ [C(g)] = 169.98±0.11 kcal/mol [69]. (The uncertainty in $\Delta H^\circ_{f,0}$ [H(g)] is negligible.)

nored. And manifestly, even a 2 % error in a 62 kcal/mol zero-point vibrational energy would be unacceptable.

Let us now consider the more approximate results. While W1h coincidentally agrees to better than 0.2 kcal/mol with the W2h result, W1 deviates from the latter by 0.6 kcal/mol. Note, however, that in W1h theory, the extrapolations bridge gaps of 0.8 (SCF), 10.1 (CCSD), and 2.1 (T) kcal/mol, the corresponding amounts for W1 theory being 0.7, 9.1, and 1.9 kcal/mol, respectively. Common sense suggests that if extrapolations account for 13.0 (W1h) and 11.7 (W1) kcal/mol, then a discrepancy of 1 kcal/mol should not come as a surprise – in fact, the relatively good agreement between the two sets of numbers and the more rigorous W2h result (total extrapolation: 6.3 kcal/mol) testifies, if anything, to the robustness of the method.

As for the difference of about 0.4 kcal/mol between the old-style and new-style SCF extrapolations in W1h and W1 theories, comparison with the W2h SCF limits clearly suggests the new-style extrapolation to be the more reliable one. (The two extrapolations yield basically the same result in W2h.) This should not be seen as an indication that the $E_\infty + A/L^5$ formula is somehow better founded theoretically, but rather as an example of why reliance on (aug-)cc-pVDZ data should be avoided if at all possible. Users who prefer the geometric extrapolation for the SCF component could consider carrying out a direct SCF calculation in the "extra large" (i.e. V5Z) basis set and applying the $E_\infty + A/B^L$ extrapolation to the "medium", "large", and "extra large" SCF data.

6. CONCLUSIONS AND PROSPECTS

W1/W2 theory and their variants would appear to represent a valuable addition to the computational chemist's toolbox, both for applications that require high-accuracy energetics for small molecules and as a potential source of parameterization data for more approximate methods. The extra cost of W2 theory (compared to W1 theory) does appear to translate into better results for heats of formation and electron affinities, but does not appear to be justified for ionization potentials and proton affinities, for which the W1 approach yields basically converged results. Explicit calculation of anharmonic zero-point energies (as opposed to scaling of harmonic ones) does lead to a further improvement in the quality of W2 heats of formation; at the W1 level, the improvement is not sufficiently noticeable to justify the extra expense and difficulty.

Of the various reduced-cost variants introduced in this paper, W2h performs basically as accurately as to W2 for heats of formation. Likewise, W1h is essentially as good as W1 theory for ionization potentials, and almost as good for heats of formation. Neither method is recommended for electron affinities.

In systems where a large number of inner-shell electrons makes the inner-shell correlation (and, to a lesser extent, scalar relativistic) steps in W1 and W2 theory unfeasible, the use of a bond equivalent model for the inner-shell correlation and scaled B3LYP/cc-pVTZuc+1 scalar relativistic corrections offers an alternative under the name of W1c and W1ch theories.

One plan for the future is the extension to heavier element systems; the first step in this direction has been made recently with the development of the SDB-cc-pVnZ valence basis sets [84] (for use with the

Stuttgart-Dresden-Bonn relativistic ECPs [85]) for third- and fourth-row main group elements.

Further improvement of accuracy, as well as applicability to systems exhibiting nondynamical correlation, will almost certainly require some level of treatment of connected quadruple excitations.

ACKNOWLEDGEMENTS

JM is the incumbent of the Helen and Milton A. Kimmelman Career Development Chair. SP acknowledges a Clore Postdoctoral Fellowship. This research was supported by the *Tashtiyot* (Infrastructures) program of the Ministry of Science and Technology (Israel). The authors would like to acknowledge helpful discussions with Prof. Peter R. Taylor (UCSD), Drs. Charles W. Bauschlicher Jr. and Timothy J. Lee (NASA Ames Research Center), Drs. David A. Dixon and Thom. H. Dunning Jr. (PNNL), Prof. Martin Head-Gordon (UC Berkeley), and finally Dr. Michael J. Frisch (Gaussian, Inc.), who originally suggested to us that we consider B3LYP as a reduced-cost alternative for the scalar relativistic correction. We thank Mark Iron for editorial assistance and Dr. Angela K. Wilson for a preprint of Ref. 12.

REFERENCES

1. J. M. L. Martin and G. de Oliveira, *J. Chem. Phys.* **111**, 1843 (1999).
2. T. H. Dunning Jr., *J. Chem. Phys.* **90**, 1007 (1989).
3. T. H. Dunning Jr., K. A. Peterson, and D. E. Woon, "Correlation consistent basis sets for molecular calculations", in *Encyclopedia of Computational Chemistry*, P. von Ragué Schleyer (Ed.), Wiley & Sons, Chichester, UK (1998).
4. J. M. L. Martin, *J. Chem. Phys.* **97**, 5012 (1992).
5. R. A. Kendall, T. H. Dunning, and R. J. Harrison, *J. Chem. Phys.* **96**, 6796 (1992).
6. J. E. Del Bene, *J. Phys. Chem.* **97**, 107 (1993).
7. J. M. L. Martin and P. R. Taylor, *Chem. Phys. Lett.* **225**, 473 (1994).
8. J. M. L. Martin, *Chem. Phys. Lett.* **242**, 343 (1995).
9. D. E. Woon and T. H. Dunning, Jr., *J. Chem. Phys.* **103**, 4572 (1995).
10. J. M. L. Martin and O. Uzan, *Chem. Phys. Lett.* **282**, 19 (1998).
11. J. M. L. Martin, *J. Chem. Phys.* **108**, 2791 (1998).
12. T. H. Dunning Jr., K. A. Peterson, and A. K. Wilson, *J. Chem. Phys.*, in press (preprint communicated to the authors).

13. A. D. Becke, *J. Chem. Phys.* **98**, 5648 (1993).
14. C. Lee, W. Yang, and R. G. Parr, *Phys. Rev.* **B 37**, 785 (1988).
15. J. M. L. Martin, J. El-Yazal, and J. P. François, *Mol. Phys.* **86**, 1437 (1995).
16. J. M. L. Martin, *Spectrochim. Acta* **A 55**, 709 (1999).
17. J. M. L. Martin, *J. Chem. Phys.* **100**, 8186 (1994).
18. J. M. L. Martin, *Chem. Phys. Lett.* **292**, 411 (1998).
19. F. Jensen, *J. Chem. Phys.* **110**, 6601 (1999).
20. F. Jensen, *Theor. Chem. Acc.* **104**, 484 (2000).
21. J. M. L. Martin and P. R. Taylor, *Mol. Phys.* **96**, 681 (1999).
22. D. Feller, *J. Chem. Phys.* **96**, 6104 (1992).
23. G. A. Petersson, A. Bennett, T. G. Tensfeldt, M. A. Al-Laham, W. A. Shirley, and J. Mantzaris, *J. Chem. Phys.* **89**, 2193 (1988).
24. M. Abramowitz and I. A. Stegun, *Handbook of mathematical functions*, Dover, New York (1972).
25. J. M. L. Martin and P. R. Taylor, *J. Chem. Phys.* **106**, 8620 (1997).
26. S. Parthiban and J. M. L. Martin, *J. Chem. Phys.* **114**, xxxx (2001).
27. C. W. Bauschlicher Jr. and H. Partridge, *Chem. Phys. Lett.* **240**, 533 (1995).
28. J. M. L. Martin, *Chem. Phys. Lett.* **310**, 271 (1999).
29. J. M. L. Martin, unpublished.
30. G. D. Purvis III and R. J. Bartlett, *J. Chem. Phys.* **76**, 1910 (1982).
31. C. Schwartz, in *Methods in Computational Physics 2* B. J. Alder (Ed.), Academic Press, New York (1963).
32. R. N. Hill, *J. Chem. Phys.* **83**, 1173 (1985).
33. W. Kutzelnigg and J. D. Morgan III, *J. Chem. Phys.* **96**, 4484 (1992); *erratum* **97**, 8821 (1992).
34. J. M. L. Martin, *Chem. Phys. Lett.* **259**, 669 (1996).
35. A. Halkier, T. Helgaker, P. Jørgensen, W. Klopper, H. Koch, J. Olsen, and A. K. Wilson, *Chem. Phys. Lett.* **286**, 243 (1998).
36. T. J. Lee and G. E. Scuseria, in *Quantum mechanical electronic structure calculations with chemical accuracy*, S. R. Langhoff (Ed.), Kluwer, Dordrecht (The Netherlands) (1995) p. 47; P. R. Taylor, in *Lecture Notes in Quantum Chemistry II*, B. O. Roos (Ed.), *Lecture Notes in Chemistry* **64**, Springer, Berlin (1994) p. 125; R. J. Bartlett and J. F. Stanton, in *Reviews in Computational Chemistry*, Vol. V, K. B. Lipkowitz and D. B. Boyd (Eds.), VCH, New York (1994) p. 65.
37. K. Raghavachari, G. W. Trucks, J. A. Pople, and M. Head-Gordon, *Chem. Phys. Lett.* **157**, 479 (1989); for alternative implementations see: A. P. Rendell, T. J. Lee and A. Komornicki, *Chem. Phys. Lett.* **178**, 462 (1991); G. E. Scuseria, *Chem. Phys. Lett.* **176**, 27 (1991); P. J. Knowles, C. Hampel, and H. J. Werner, *J. Chem. Phys.* **99**, 5219 (1993); *erratum* **112**, 3106 (2000); J. D. Watts, J. Gauss, and R. J. Bartlett, *J. Chem. Phys.* **98**, 8718 (1993).
38. Y. He, Z. He, and D. Cremer, *Theor. Chem. Acc.* **105**, 182 (2001).
39. W. Klopper, J. Noga, H. Koch, and T. Helgaker, *Theor. Chem. Acc.* **97**, 164 (1997).

40. H. Koch, A. Sanchez de Meras, T. Helgaker, and O. Christiansen, *J. Chem. Phys.* **104**, 4157 (1996) and subsequent papers.

41. M. Schütz, R. Lindh, and H.-J. Werner, *Mol. Phys.* **96**, 719 (1999).

42. C. W. Bauschlicher Jr., S. R. Langhoff, and P. R. Taylor, *J. Chem. Phys.* **88**, 2540 (1988).

43. J. M. L. Martin, A. Sundermann, P. L. Fast, and D. G. Truhlar, *J. Chem. Phys.* **113**, 1348 (2000).

44. P. Pyykkö, *Chem. Rev.* **88**, 563 (1988); M. Reiher and B. A. Hess, in *Modern methods and algorithms of quantum chemistry*, J. Grotendorst (Ed.) NIC Series Vol. 1, Forschungszentrum Jülich (2000).

45. G. de Oliveira, J. M. L. Martin, F. de Proft, and P. Geerlings, *Phys. Rev.* **A 60**, 1034 (1999).

46. R. D. Cowan and M. Griffin, *J. Opt. Soc. Am.* **66**, 1010 (1976).

47. R.L. Martin *J. Phys. Chem.* **87**, 750 (1983).

48. C. W. Bauschlicher Jr., *J. Phys. Chem.* **A 104**, 2281 (2000).

49. C. W. Bauschlicher Jr., *Theor. Chem. Acc.* **101**, 421 (1999).

50. A. Nicklass, K. A. Peterson, A. Berning, H.-J. Werner, and P. J. Knowles, *J. Chem. Phys.* **112**, 5624 (2000).

51. R. S. Grev, C. L. Janssen, and H. F. Schaefer III, *J. Chem. Phys.* **95**, 5128 (1991).

52. A. P. Scott and L. Radom, *J. Phys. Chem.* **100**, 16502 (1996).

53. D. Feller and D. A. Dixon, *J. Phys. Chem.* **A 104**, 3048 (2000).

54. J. M. L. Martin, T. J. Lee, P. R. Taylor, and J. P. François, *J. Chem. Phys.* **103**, 2589 (1995); J. M. L. Martin and P. R. Taylor, *Chem. Phys. Lett.* **248**, 336 (1996).

55. E. Miani, E. Cané, P. Palmieri, A. Trombetti, and N. C. Handy, *J. Chem. Phys.* **112**, 248 (2000).

56. J. M. L. Martin, P. R. Taylor, and T. J. Lee, *Chem. Phys. Lett.* **275**, 414 (1997).

57. G. de Oliveira, J. M. L. Martin, I. K. C. Silwal, and J. F. Liebman, *J. Comput. Chem.*, in press (2001) [Paul von Ragué Schleyer festschrift].

58. A. D. Becke, *J. Chem. Phys.* **98**, 1372 (1993).

59. B. J. Lynch, P. L. Fast, M. Harris, and D. G. Truhlar, *J. Phys. Chem.* **A 104**, 4811 (2000).

60. A. Nicklass and K. A. Peterson, *Theor. Chem. Acc.* **100**, 103 (1998).

61. W. Müller, J. Flesch, and W. Meyer, *J. Chem. Phys.* **80**, 3297 (1984); P. Fuentealba, H. Preuss, H. Stoll, and L. von Szentpály, *Chem. Phys. Lett.* **89**, 418 (1982). P. Schwerdtfeger and H. Silberbach, *Phys. Rev.* **A 37**, 2834 (1988); *erratum* **42**, 665 (1990).

62. G. S. Kedziora, J. A. Pople, V. A. Rassolov, M. A. Ratner, P. C. Redfern, and L. A. Curtiss, *J. Chem. Phys.* **110**, 7123 (1999).

63. Supplementarty material to the present paper is available at the URL `http://theochem.weizmann.ac.il/web/papers/w1chapter.html`.

64. T. J. Lee and P. R. Taylor, *Int. J. Quantum Chem. Symp.* **23**, 199 (1989); for a generalization to open-shell cases, see D. Jayatilaka and T. J. Lee, *J. Chem. Phys* **98**, 9734 (1993).
65. M. R. Hoffmann and H. F. Schaefer III, *Adv. Quantum Chem.* **18**, 207 (1986); J. Noga and R. J. Bartlett, *J. Chem. Phys* **86**, 7041 (1987); *erratum* **89**, 3401 (1988); G. E. Scuseria and H. F. Schaefer III, *Chem. Phys. Lett.* **152**, 382 (1988).
66. K. L. Bak, P. Jørgensen, J. Olsen, T. Helgaker, and J. Gauss, *Chem. Phys. Lett.* **317**, 116 (2000).
67. J. M. L. Martin, *Chem. Phys. Lett.* **303**, 399 (1999).
68. J. M. L. Martin *Spectrochim. Acta* **A 57**, 875 (2001) [special issue on astrophysically important molecules].
69. J.D. Cox, D.D. Wagman, and V.A. Medvedev, *CODATA key values for thermodynamics* (Hemisphere, New York, 1989). [Data also available online at `http://www.codata.org/codata/databases/key1.html`]
70. G. K. Johnson, *J. Chem. Thermodyn.* **18**, 801 (1986).
71. K. P. Huber and G. Herzberg, *Constants of diatomic molecules* Van Nostrand Reinhold, New York (1979).
72. C. W. Bauschlicher Jr., J. M. L. Martin, and P. R. Taylor, *J. Phys. Chem.* **A 103**, 7715 (1999); see also J. M. L. Martin, and P. R. Taylor, *J. Phys. Chem.* **A 102**, 2995 (1998).
73. J. M. L. Martin, and P. R. Taylor, *J. Phys. Chem.* **A 103**, 4427 (1999).
74. D. F. Hildenbrand, personal communication quoted in Ref. 72.
75. E. Storms and B. Mueller, *J. Phys. Chem.* **81**, 318 (1977).
76. J.A. Ochterski, G.A. Petersson, and K.B. Wiberg, *J. Am. Chem. Soc.* **117**, 11299 (1995).
77. J. Baker, M. Muir, and J. Andzelm, *J. Chem. Phys.* **102**, 2063 (1995).
78. J. Andzelm and P. R. Taylor, *Chem. Phys. Lett.* **237**, 53 (1995).
79. C. Adamo and V. Barone, *J. Chem. Phys.* **108**, 664 (1998).
80. S. Parthiban, G. de Oliveira, and J. M. L. Martin, *J. Phys. Chem.* **A 105**, 895 (2001).
81. A. D. Becke, *Phys. Rev.* **A 38**, 3098 (1988); *J. Chem. Phys.* **88**, 2547 (1988).
82. A. D. Boese, N. L. Doltsinis, N. C. Handy, and M. Sprik, *J. Chem. Phys.* **112**, 1670 (2000); see also F. A. Hamprecht, A. J. Cohen, D. J. Tozer, and N. C. Handy, *J. Chem. Phys.* **109**, 6264 (1998).
83. M. N. Glukhovtsev, A. Pross, and L. Radom, *J. Am. Chem. Soc.* **117**, 2024 (1995); *J. Am. Chem. Soc.* **118**, 6273 (1996).
84. J. M. L. Martin and A. Sundermann, *J. Chem. Phys.* **114**, 3408 (2001).
85. For a review see M. Dolg, in *Modern methods and algorithms of quantum chemistry*, J. Grotendorst (Ed.), NIC Series Vol. 1, John von Neumann-Institute for Computing, Forschungszentrum Jülich, Germany (2000).

Chapter 3

Quantum-Chemical Methods for Accurate Theoretical Thermochemistry

Krishnan Raghavachari
Electronic and Photonic Materials Physics Research, Agere Systems, Murray Hill, NJ 07974, U.S.A.

Larry A. Curtiss
Materials Science and Chemistry Divisions, Argonne National Laboratory, Argonne, IL 60439, U.S.A.

1. INTRODUCTION

The first-principles evaluation of the binding energies of molecules to chemical accuracy (± 1 kcal/mol) is one of the most challenging problems in computational quantum chemistry. Dramatic progress has been made in this regard in the last two decades and the rigorous demands placed on the theoretical methods to achieve this goal are now well understood. In principle it is now known how to compute the binding energies and other thermochemical properties of most molecules to very high accuracy [1-10]. This can be achieved by using very high levels of correlation, such as that obtained with coupled cluster [CCSD(T)] [11] or quadratic configuration interaction [QCISD(T)] [12] methods, and very large basis sets containing high angular momentum functions. The results of these calculations are then extrapolated to the complete basis set limit and corrected for some smaller effects such as core-valence and relativistic effects. Unfortunately, this approach is limited to small molecules because of the $\sim N^7$ scaling (with respect to the number of

J. Cioslowski (ed.), Quantum-Mechanical Prediction of Thermochemical Data, 67–98.

basis functions N) of the correlation methods and the need for very large basis sets.

An alternative approach applicable for larger molecules is to use a series of high-level correlation calculations [e.g., QCISD(T), MP4 [13], or CCSD(T)] with moderate sized basis sets to approximate the result of a more expensive calculation. The Gaussian-n series [14-30] exploits this idea to predict thermochemical data. In addition, molecule-independent empirical parameters are used in these methods to estimate the remaining deficiencies in the calculations. Such an approach using higher-level corrections (additive parameters that depend on the number of paired and unpaired electrons in the system) has been quite successful, and the latest version, Gaussian-3 (G3) theory [21], achieves an overall accuracy of 1 kcal/mol for the G2/97 test set [24, 25]. Petersson et al. [31] have developed a related series of methods, referred to as complete basis set (CBS) procedures, for the evaluation of accurate energies of molecular systems. The central idea in the CBS methods is an extrapolation procedure to determine the projected second-order (MP2) energy in the limit of a complete basis set. Several empirical corrections, similar in spirit to the higher-level correction used in the Gaussian-n series, are added to the resulting energies in the CBS methods to remove systematic errors in the calculations. Another approach to calculation of thermochemical data that has been proposed is scaling of the calculated correlation energy using multiplicative parameters [32-36] determined by fitting to experimental data. Finally, hybrid density functionals are being used increasingly to predict the thermochemistry of molecules with reasonable accuracy [24-26].

In this chapter, we review the elements of G3 theory and related techniques of computational thermochemistry. This review is restricted almost exclusively to the techniques that we have developed and the reader is referred to the remaining chapters in this volume for other complementary approaches. An important part of the development of such quantum-chemical methods is their critical assessment on test sets of accurate experimental data. Section 3.2 provides a brief description of the comprehensive G3/99 test set [26] of experimental data that we have collected. Section 3.3 discusses the components of G3 theory as well as the approximate versions such as G3(MP3) [22] and G3(MP2) [23], and their performance for the G3/99 test set. The G3S method [29] that includes multiplicative scale factors is presented in section 3.4 along with other related variants. Section 3.5 discusses the recently developed G3X method [30] that corrects for most of the deficiencies of G3 theory for larger molecules. The performance of these methods is compared to that of some of the popularly used density functionals in section 3.6. Finally, conclusions are drawn in section 3.7.

2. THE G3/99 TEST SET

Critical documentation and assessment of quantum-chemical models is essential for such methods to become predictive tools for chemical investigation. We have assembled a large test set of good, credible experimental data to perform such assessments [24-26]. The current test set, referred to as G3/99 [26], contains 376 energies (222 enthalpies of formation, 88 ionization energies, 58 electron affinities, and 8 proton affinities) that are known experimentally [37-39] to an accuracy of better than ±1 kcal/mol. It includes three subsets of energies, G2-1, G2-2, and G3-3. The G2-1 subset (original G2 test set) includes the enthalpies of formation for only very small molecules containing 1 - 3 heavy atoms (systems such as H_2O, C_2H_4, CO_2 and SO_2), whereas G2-2 includes medium-sized molecules containing 3 - 6 heavy atoms (systems such as C_3H_6, C_4H_4O, C_6H_6, etc.). It also includes ionization energies on some larger molecules such as substituted benzenes. The two subsets, G2-1 and G2-2, are together referred to as G2/97 and contain 301 test energies [24, 25]. The G3-3 test set [26] comprises 75 new enthalpies of formation for molecules that are, on average, larger (containing 3 - 10 heavy atoms). The largest molecule in the G3-3 test set contains ten non-hydrogen atoms (e.g., naphthalene or azulene). It also includes some larger hypervalent molecules such as PF_5 or SF_6 that provide a challenge for many theoretical models.

The 222 enthalpies of formation included in the G3/99 test set contain a wide variety of molecules with many different kinds of bonds. They are conveniently classified into subgroups of molecules. They include 47 molecules containing non-hydrogen atoms, 38 hydrocarbons, 91 substituted hydrocarbons, 15 inorganic hydrides, and 31 open-shell radicals. Together, they provide a comprehensive assessment of new theoretical models in a wide variety of bonding environments.

The collection of such a large set of experimental data provides many challenges. All the experimental values that are included have a quoted uncertainty of less than 1 kcal/mol [37-39]. However, the evaluation of the experimental uncertainties is difficult or impossible in many cases. It is possible that some of the included values may turn out to be incorrect. For example, the G2/97 test set originally comprised 302 energies, but the enthalpy of formation of COF_2 has been deleted because a new experimental upper limit [40] has been reported that casts doubt on the value used in the G2/97 test set. In addition, on the basis of theoretical evidence, a few other enthalpies of formation (vinyl chloride, C_2F_4, and C_2Cl_4) and one ionization energy (B_2F_4) may not be as accurate as cited experimentally [41,42]. In our analysis, we have chosen not to throw out experimental data unless there is new *experimental* evidence that warrants it. Another important factor is that the calculation of the enthalpies of formation for molecules requires the experimental atomic enthalpies of formation. Two of these (B and Si) have significant uncertainties and some authors have suggested the use of "theoretical" atomic enthalpies of formation for Si and B in the calculation of molecular enthalpies of formation [43-45]. We have consistently used experimental values for all elements, despite the uncertainty in the Si and B values. The reason that we do not use these "theoretical" atomic enthalpies is that they are derived in part from an experimental molecular enthalpy that is part of the test set, which may bias the assessment process [46]. If the accuracy of theory improves and becomes demonstrably better than that of experiment, theoretical values may be included in the future to assemble test sets of molecules for critical assessment.

3. GAUSSIAN-3 THEORY

Gaussian-3 theory, like its predecessor Gaussian-2 (G2) theory [17], is a composite technique in which a sequence of well-defined ab initio molecular orbital calculations [47] is performed to arrive at a total energy of a given molecular species. It was designed to correct some of the deficiencies of G2 theory for systems such as halogen-containing molecules, unsaturated hydrocarbons, etc. It also contains important physical effects, such as core-valence correlation and spin-orbit contributions, that were not included in G2 theory. G3 theory is computationally less demanding than G2 theory though it is significantly more accurate. The detailed steps involved in G3 theory are as follows:

1. An initial equilibrium structure is obtained at the Hartree-Fock (HF) level with the 6-31G(d) basis [47]. Spin-restricted (RHF) theory is used for singlet states and spin-unrestricted Hartree-Fock theory (UHF) for others. The HF/6-31G(d) equilibrium structure is used to calculate harmonic frequencies, which are then scaled by a factor of 0.8929 to take account of known deficiencies at this level [48]. These frequencies are used to evaluate the zero-point energy E_{ZPE} and thermal effects.

2. The equilibrium geometry is refined at the MP2(fu)/6-31G(d) level, using all electrons for the calculation of correlation energies. This is the final equilibrium geometry in the theory and is used for all single-point calculations at higher levels of theory in step 3. Except where otherwise noted by the symbol (fu), these subsequent calculations include only valence electrons in the treatment of electron correlation.

3. A series of single-point energy calculations is carried out at higher levels of theory. The first higher-level calculation is the complete fourth-order Møller-Plesset perturbation theory [13] with the 6-31G(d) basis set, i.e. MP4/6-31G(d). For convenience of notation, we represent this as MP4/d. This energy is then modified by a series of corrections from additional calculations:

   (a) A correction E_{QCI} for correlation effects beyond fourth-order perturbation theory using the quadratic configuration interaction (QCI) method [12],

$$E_{QCI} = E_{QCISD(T)/d} - E_{MP4/d}\,. \tag{3.1}$$

   (b) A correction E_{plus} for diffuse functions,

$$E_{plus} = E_{MP4/plus} - E_{MP4/d}\,, \tag{3.2}$$

   where plus denotes the 6-31+G(d) basis set [47].

   (c) A correction $E_{2df,p}$ for higher polarization functions on non-hydrogen atoms and p-functions on hydrogens,

$$E_{2df,p} = E_{MP4/2df,p} - E_{MP4/d}\,, \tag{3.3}$$

   where 2df,p denotes the polarized 6-31G(2df,p) basis set [47].

   (d) A correction $E_{G3Large}$ for larger basis set effects and for the non-additivity caused by the assumption of separate basis set extensions for diffuse functions and higher polarization functions,

$$E_{G3Large} = E_{MP2(fu)/G3Large} - E_{MP2/2df,p} - E_{MP2/plus} + E_{MP2/d}\,. \tag{3.4}$$

   The largest basis set, denoted as G3Large [21] includes some core polarization functions as well as multiple sets of valence polarization functions. It should be noted that MP2 calculation with the largest basis set in Eq. (3.4) is carried out at the MP2(fu) level.

This is done to take account of core-related correlation contributions to total energies.

4. Spin-orbit correction E_{SO} is included for atomic species *only*. The spin-orbit correction is taken from experiment [49] where available and accurate theoretical calculations [50] in other cases. These corrections are particularly important for halide-containing systems [24]. Molecular spin-orbit corrections are not included in G3 theory.

5. A "higher-level correction" E_{HLC} is added to take into account remaining deficiencies in the energy calculations: E_{HLC} is given by $-An_\beta - B(n_\alpha - n_\beta)$ for molecules and $-Cn_\beta - D(n_\alpha - n_\beta)$ for atoms (including atomic ions), where the n_β and n_α are the numbers of β and α valence electrons, respectively, with $n_\alpha \geq n_\beta$. The number of valence electron pairs corresponds to n_β. Thus, A is the correction for pairs of valence electrons in molecules, B is the correction for unpaired electrons in molecules, C is the correction for pairs of valence electrons in atoms, and D is the correction for unpaired electrons in atoms. The use of different corrections for atoms and molecules can be justified, in part, by noting that effects of basis functions with higher angular momentum are likely to be of more importance in molecules than in atoms. The A, B, C, D values are chosen to give the smallest average absolute deviation from experiment for the G2/97 test set. For G3 theory, A = 6.386 mE_h, B = 2.977 mE_h, C = 6.219 mE_h, D = 1.185 mE_h.

6. Finally, the total energy at 0 K ("G3 energy") is obtained by adding all the individual energy corrections in an additive manner,

$$\begin{aligned} E^o_{G3} &= E_{MP4/d} + E_{QCI} + E_{plus} + E_{2df,p} \\ &+ E_{G3Large} + E_{SO} + E_{HLC} + E_{ZPE}\,. \end{aligned} \tag{3.5}$$

The G3 energy can also be represented more fully as

$$\begin{aligned} E^o_{G3} &= E_{MP4/d} + (E_{QCISD(T)/d} - E_{MP4/d}) \\ &+ (E_{MP4/plus} - E_{MP4/d}) + (E_{MP4/2df,p} - E_{MP4/d}) \\ &+ (E_{MP2(fu)/G3Large} - E_{MP2/2df,p} - E_{MP2/plus} + E_{MP2/d}) \\ &+ E_{SO} + E_{HLC} + E_{ZPE}\,. \end{aligned} \tag{3.6}$$

The final total energy is effectively at the QCISD(T,fu)/G3Large level if the additivity approximations used work well. The validity of such approximations has been previously investigated for G2 theory on the G2-1 subset of G2/97 and found to be satisfactory [19].

The correlation methods in G3 theory are still computationally demanding and it is of interest to find modifications to reduce the computational requirements. Two approximate versions of G3 theory have been proposed to make the methods more widely applicable. The first is G3(MP3) [22] that eliminates the expensive MP4/2df,p calculation by evaluating the larger basis set effects at the MP3 level. It also eliminates the MP4/plus calculation,

$$\begin{aligned} E^{o}_{G3(MP3)} &= E_{MP4/d} + (E_{QCISD(T)/d} - E_{MP4/d}) \\ &+ (E_{MP3/2df,p} - E_{MP3/d}) + (E_{MP2(fu)/G3Large} - E_{MP2/2df,p}) \\ &+ E_{SO} + E_{HLC} + E_{ZPE}\,. \end{aligned} \tag{3.7}$$

The second is G3(MP2) theory [23] that evaluates the larger basis set effects at the MP2 level, similar to the successful G2(MP2) theory,

$$\begin{aligned} E^{o}_{G3(MP2)} &= E_{MP4/d} + (E_{QCISD(T)/d} - E_{MP4/d}) \\ &+ (E_{MP2/G3MP2Large} - E_{MP2/d}) \\ &+ E_{SO} + E_{HLC} + E_{ZPE}\,. \end{aligned} \tag{3.8}$$

In G3(MP2) theory, the MP2(fu)/G2Large calculation of G3 is replaced with a frozen core calculation with the G3MP2Large basis set [23] that does not contain the core polarization functions of the G3Large basis set.

The enthalpies of formation for most molecules in the G2/97 and G3/99 test sets have been measured at 298 K. In order to compare with experiment, the heats of formation for molecules are calculated using a procedure described in detail previously [24]. Briefly, thermal corrections (298 K) are first evaluated using the calculated vibrational frequencies and standard statistical-mechanical methods [51]. The calculated total energies of the given molecule and its constituent atoms are used to evaluate its atomization energy. This value is then used along with the thermal corrections and the known experimental enthalpies of formation for the atomic species [21, 38] to calculate the enthalpy of formation for the molecule (298 K). The electron affinities are calculated as the difference in total energies at 0 K of the anion and the corresponding neutral at their respective MP2(fu)/6-31G(d) optimized geometries. Likewise, the ionization potentials are calculated as the difference in total energies at 0 K of the cation and the corresponding neutral at their respective MP2(fu)/6-31G(d) optimized geometries. The Gaussian 98 computer

Table 3.1 Summary of mean absolute deviations (kcal/mol) for G3 theories.

Set of data[a]	G3	G3(MP3)	G3(MP2)
The G2/97 test set			
Enthalpies of formation (147)	0.92	1.19	1.17
Non-hydrogens (34)	1.68	2.09	2.06
Hydrocarbons (22)	0.68	0.86	0.70
Subst. hydrocarbons (47)	0.56	0.78	0.74
Inorganic hydrides (15)	0.87	1.18	1.03
Radicals (29)	0.84	1.05	1.23
All (301)[b]	1.01	1.21	1.31
The complete G3/99 set			
Enthalpies of formation (222)	1.05	1.29	1.22
Non-hydrogens (47)	2.11	2.74	2.45
Hydrocarbons (38)	0.69	0.77	0.71
Subst. hydrocarbons (91)	0.75	0.86	0.83
Inorganic hydrides (15)	0.87	1.18	1.03
Radicals (31)	0.87	1.06	1.21
Ionization energies (88)	1.14	1.24	1.46
Electron affinities (58)	0.98	1.24	1.46
Proton affinities (8)	1.34	1.25	1.02
All (376)[b]	1.07	1.27	1.31

[a] Number of species in each set given in parentheses.

[b] The mean absolute deviations for the ionization energies, electron affinities, and proton affinities in the G2/97 test set are the same as in G3/99.

program is used for the calculations [52]. Many of the G3 techniques have been implemented in this computer program.

The performance of G3, G3(MP3), and G3(MP2) theories for the energies in the G2/97 and G3/99 test sets is summarized in Table 3.1. Overall, the mean absolute deviations increase slightly for the G3/99 test set compared to that of the G2/97 test set. The mean absolute deviation of G3 theory increases from 1.01 kcal/mol to 1.07 kcal, that of G3(MP3) theory increases from 1.21 kcal/mol to 1.27 kcal/mol, and that of G3(MP2) theory remains at 1.31 kcal/mol. This increase in the mean absolute deviation is primarily due to large deviations in the calculated enthalpies of formation of some of the non-hydrogen species in the expanded test set. In particular, the mean absolute deviation of

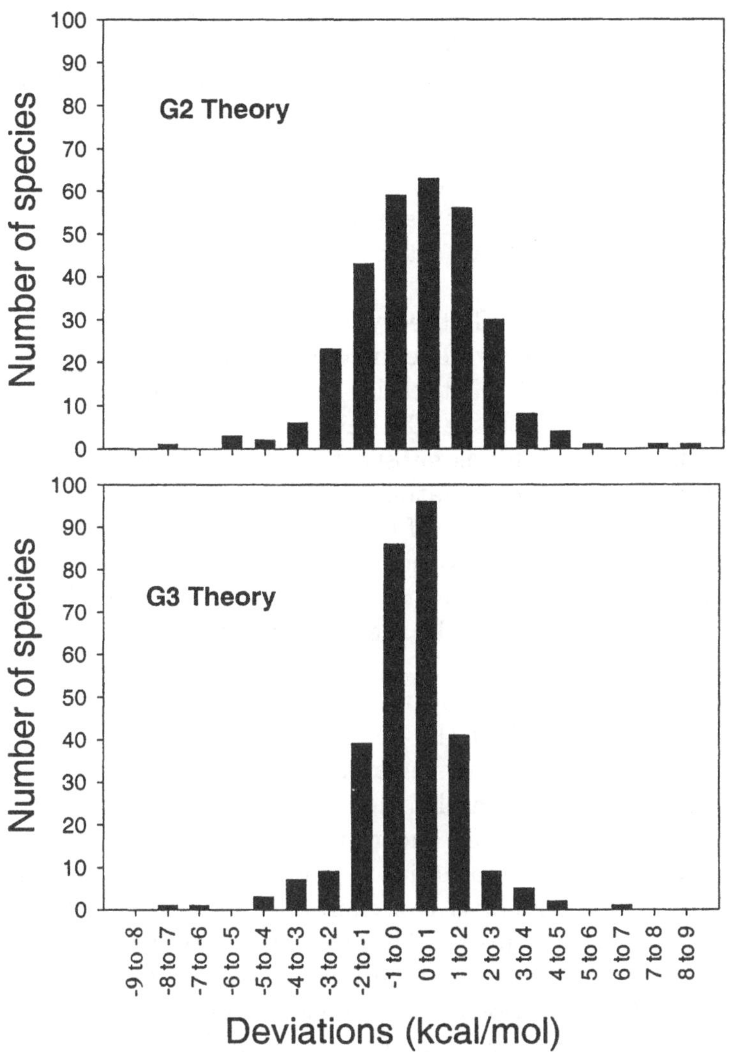

Figure 3.1 Histograms of G2 and G3 deviations for the G2/97 test set. Each vertical bar represents deviations (experiment - theory) of 1 kcal/mol range.

3.24 kcal/mol for the 13 non-hydrogens in the G3-3 subset is nearly twice that of the 34 non-hydrogens in the G2/97 set (1.68 kcal/mol). Especially large deviations in the G3 data occur for SF_6 (-6.22 kcal/mol), PF_5 (-7.05 kcal/mol), SO_3 (-5.14 kcal/mol), P_4 (-4.15 kcal/mol), and SO_2Cl_2 (-4.37 kcal/mol). Among these, P_4 is an unusually strained molecule with a bond angle of 60°. The remaining systems are hypervalent. Overall, the G3 deviations for nearly all of the new non-hydrogen species are negative, indicating underbinding. Part of the error for these species is due to the use of MP2/6-31G(d) geometries. For example, when experimental geometries are used, the deviation for SF_6 decreases from -6.22 to -3.42 kcal/mol, that for PF_5 decreases from -7.05 to -4.93 kcal/mol, and that for SO_3 falls from -5.14 to -2.53 kcal/mol. The remainder of the discrepancies for non-hydrogen systems is mostly due to basis set deficiencies. The mean absolute deviations for the other types of molecules in the G3/99 test set are similar to those in the G2/97 test set.

As mentioned earlier, G3 theory was designed to correct for some of the deficiencies in G2 theory. The histograms in Fig. 3.1 show the range of deviations of G2 and G3 theories from experiment for the G2/97 test set. Nearly 88 % of the G3 deviations fall within the range of -2.0 to 2.0 kcal/mol. This is substantially better than G2 theory for which about 74 % of the deviations fall in this range. In addition to improving the accuracy, the use of the 6-31G(d)-based calculations in G3 theory substantially decreases the computer time as well as disk space requirements relative to G2 theory [which uses the larger 6-311G(d,p)-based calculations]. For example, the G3 calculation on benzene is nearly twice as fast as the analogous G2 calculation.

As proposed originally, G3 theory is applicable only to molecules containing atoms of the first (Li - F) and second (Na - Cl) rows of the periodic chart. It has recently been extended [53] to molecules containing the third-row non-transition elements K, Ca, and Ga - Kr. Basis sets compatible to those used in G3 theory for molecules containing first- and second-row atoms have been derived. The G3 mean absolute deviation from experiment for a set of 47 test energies containing these elements is 0.94 kcal/mol. This is a substantial improvement over G2 theory for the third row, which has a mean absolute deviation of 1.43 kcal/mol for the same set [54, 55]. Variations of G3 theory based on reduced orders of perturbation theory that are similar to those for G2 theory [56] have also been reported [53]. G3(MP2) theory for third-row molecules has a mean absolute deviation from experiment of 1.30 kcal/mol, and is significantly more accurate than G2(MP2). The G3 method based on third-order perturbation theory, G3(MP3), has an average absolute deviation of 1.24 kcal/mol. In addition, these methods have been assessed on a set of molecules containing K and Ca for which the experimental data is not accurate enough for them to be included in the test set [53]. Results for this set indicate that G3 theory performs significantly better than G2 for molecules containing Ca.

Other variants of G3 theory have been proposed that use alternate geometries, zero-point energies, or higher-order correlation methods. G3 theory uses MP2(fu)/6-31G(d) geometries, and scaled HF/6-31G(d) frequencies and zero-point energies. A method using B3LYP/6-31G(d) geometries and scaled B3LYP/6-31G(d) zero-point energies (0.96) has been considered to make it more uniform. Denoted as G3//B3LYP [27], its performance is very similar to that of G3 theory though it may be useful in cases where the MP2 theory is deficient for geometries. Another variation involves the use of the CCSD(T)/6-31G(d) method instead of QCISD(T)/6-31G(d) to evaluate the contribution of higher-order correlation effects. The resulting G3(CCSD) method [28] has an accuracy very similar to that of G3 theory and may be useful in cases where the QCISD(T) method is not available or deficient.

In addition to these minor variants, two major variants (G3S and G3X) [29, 30] have been proposed to address some of the main deficiencies of G3 theory. These are discussed in detail in the next two sections of this chapter.

4. G3S THEORY

G3 theory and its variants discussed thus far include a higher-level correction term (HLC) to correct for the remaining deficiencies that result from basis set incompleteness, etc.. The HLC term in G3 theory consists of four molecule-independent additive parameters that depend only on the number of paired and unpaired electrons in the system. Such an approach will work if such deficiencies are systematic and scale as the number of electrons. The parameters in G3 theory were obtained by minimizing the mean absolute deviation from experiment of the energies in the G2/97 test set. This approach is indeed successful as indicated by the overall accuracy of 1 kcal/mol for this test set. However, one of the deficiencies of G3 theory is that the HLC parameters do not depend on the geometry and thus do not vary on the potential energy surface. This may cause deficiencies for regions near transition states that contain partially broken bonds. Even more importantly, G3 theory cannot be used to study potential energy surfaces for reactions where the reactants and products have a different number of electron pairs.

An alternative approach to the calculation of accurate thermochemical data is to scale the computed correlation energy with multiplicative parameters determined by fitting to the experimental data. Pioneering methods using such an approach include the scaling all correlation (SAC) method of Gordon and Truhlar [32], the parameterized correlation (PCI-X) method of Siegbahn et al. [33], and the multi-coefficient correlation methods (MCCM) of Truhlar et al. [34-36]. Such methods can be used to yield continuous potential energy surfaces even for reactions where the reactants and products contain different numbers of electron pairs.

A new family of methods, referred to as G3S (G3 Scaled), has been developed recently [29], where the additive higher-level correction is replaced by a multiplicative scaling of the correlation and Hartree-Fock components of the G3 energy. The scale factors have been obtained by fitting to the G2/97 test set of energies. This test set is substantially larger than that used in previous fits and can provide a reliable assessment of the use of such a scaling approach to computational thermochemistry.

Traditionally, the G3 energy is written in terms of corrections (basis set extensions and correlation energy contributions) to the MP4/d energy. Alternatively, the G3 energy can be specified in terms of HF and perturbation energy components. Denoting the second-, third-, and fourth-order contributions from perturbation theory by E_2, E_3, and E_4, respectively, and the contributions beyond fourth order in a QCISD(T) calculation by $E_{\Delta QCI}$, the G3 energy can be expressed as

$$\begin{aligned} E^{o}_{G3} &= E_{HF/d} + (E_{HF/G3Large} - E_{HF/d}) \\ &+ E_{2/d} + E_{3/d} + E_{4/d} + E_{\Delta QCI/d} + (E_{2(fu)/G3Large} - E_{2/d}) \\ &+ (E_{3/plus} - E_{3/d}) + (E_{3/2df,p} - E_{3/d}) + (E_{4/plus} - E_{4/d}) \\ &+ (E_{4/2df,p} - E_{4/d}) + E_{SO} + E_{HLC} + E_{ZPE}, \end{aligned} \tag{4.1}$$

where the abbreviations for other energy components are the same as in Eqs. (3.1) - (3.8). In the derivation of the scaled methods, the HLC term is set to zero and parameters are introduced that scale the different terms in the energy expression. A systematic study has been carried out to investigate the performance of different scaled methods as the number of parameters is increased. In each case, the parameters have been optimized to give the smallest root mean square deviation from experiment for the energies in the G2/97 test set.

The simplest scaled scheme can be obtained by using a single parameter to scale all the correlation energy terms in Eq. (4.1). Such a single-parameter scaling of G3 theory is similar to the SAC method of Truhlar et al. [32] and the PCI-X method of Siegbahn et al. [33]. Such a method gives a mean absolute deviation of 1.43 kcal/mol for the energies in the G2/97 test set (compared to 1.01 kcal/mol for G3 theory). On the other extreme, scaling of all 11 terms in Eq. (4.1) yields a method with a mean absolute deviation of only 0.97 kcal/mol. However, most of this improvement is obtained on using only six parameters (mean absolute deviation of 0.99 kcal/mol). Such a method is referred to as G3S theory.

The G3S energy expression is given by

$$\begin{aligned} E^o_{G3S} &= E_{HF/d} + S_{E234}\,(E_{2/d} + E_{3/d} + E_{4/d}) \\ &+ S_{QCI}\,E_{\Delta QCI/d} + S_{HF'}\,(E_{HF/G3Large} - E_{HF/d}) \\ &+ S_{E2'}\,(E_{2(fu)/G3Large} - E_{2/d}) \\ &+ S_{E3'}\,[(E_{3/plus} - E_{3/d}) + (E_{3/2df,p} - E_{3/d})] \\ &+ S_{E4'}\,[(E_{4/plus} - E_{4/d}) + (E_{4/2df,p} - E_{4/d})] \\ &+ E_{SO} + E_{ZPE}\,. \end{aligned} \tag{4.2}$$

The scale factors for the basis set extension terms ($S_{HF'}$, $S_{E2'}$, $S_{E3'}$, and $S_{E4'}$) are denoted by primes, the scale factor for the second-, third-, and fourth-order perturbation terms at the 6-31G(d) level is denoted by S_{E234}, and the scale factor for the QCI correction beyond MP4 at the 6-31G(d) level is denoted by S_{QCI}. Optimization of all six parameters in Eq. (4.2) gives a mean absolute deviation of 0.99 kcal/mol, which is slightly better than standard G3 theory with the HLC correction (mean absolute deviation of 1.01 kcal/mol). The optimized values for the parameters in the six-parameter fit are all of reasonable magnitude and range from 0.95 to 1.38. The largest scale factor occurs for the basis set extensions at the third order of perturbation theory. Only one scale factor is less than unity – that for the basis set extensions at the fourth-order perturbation theory (0.95). Thus, it is possible to obtain a very accurate version of G3 theory with scaling of energies when the basis set extensions are included in the fitting procedure.

In a similar manner, the approximate G3(MP3) method can be modified to use multiplicative scale factors. The resulting G3S(MP3) energy expression is

$$\begin{aligned} E^o_{G3S(MP3)} &= E_{HF/d} + S_{E234}\,(E_{2/d} + E_{3/d} + E_{4/d}) \\ &+ S_{QCI}\,E_{\Delta QCI/d} + S_{HF'}\,(E_{HF/G3Large} - E_{HF/d}) \\ &+ S_{E2'}\,(E_{2(fu)/G3Large} - E_{2/d}) \\ &+ S_{E3'}\,(E_{3/2df,p} - E_{3/d}) + E_{SO} + E_{ZPE}\,. \end{aligned} \tag{4.3}$$

Eq. (4.3) contains five parameters and yields a mean absolute deviation of 1.16 kcal/mol for the energies in the G2/97 test set [compared to the corresponding G3(MP3) deviation of 1.22 kcal/mol].

Finally, the G3(MP2) method can also be modified to employ multiplicative scale factors. The resulting G3S(MP2) energy expression is

Table 3.2 Summary of mean absolute deviations (kcal/mol) for G3S theories applied to the G3/99 test set.

Set of data[a]	G3S	G3S(MP3)	G3S(MP2)
Enthalpies of formation (222)	1.12	1.19	1.29
Non-hydrogens (47)	2.09	2.49	2.37
Hydrocarbons (38)	0.79	0.98	0.86
Subst. hydrocarbons (91)	0.92	0.75	0.99
Inorganic hydrides (15)	0.63	0.79	1.06
Radicals (31)	0.86	0.96	1.17
Ionization energies (88)	1.09	1.27	1.54
Electron affinities (58)	0.90	1.24	1.56
Proton affinities (8)	1.17	1.10	0.74
All (376)	1.08	1.21	1.38

[a] Number of species in each set given in parentheses.

$$\begin{aligned} E^{o}_{G3S(MP2)} &= S_{HF}\, E_{HF/d} + S_{E2}\, E_{2/d} \\ &+ S_{E34}\,(E_{3/d} + E_{4/d}) + S_{QCI}\, E_{\Delta QCI/d} \\ &+ S_{HF'}\,(E_{HF/G3MP2Large} - E_{HF/d}) \\ &+ S_{E2'}\,(E_{2/G3MP2Large} - E_{2/d}) + E_{SO} + E_{ZPE}\,. \end{aligned} \tag{4.4}$$

However, the nature of the scaling parameters in this case is somewhat different. In particular, the addition of scale factors to the $E_{HF/d}$ and $E_{2/d}$ terms was found to be important to yield good results. The resulting six-parameter fit yields a mean absolute deviation of 1.35 kcal/mol for the G2/97 test set, only slightly larger than the 1.30 kcal/mol for G3(MP2).

A summary of the mean absolute deviations of the G3S, G3S(MP3), and G3S(MP2) theories is given in Table 3.2 for the entire G3/99 test set. As mentioned earlier, the scale factors in the methods were derived from fits to the smaller G2/97 test set. For all three methods, the mean absolute deviations increase slightly for the G3/99 test set compared to its G2/97 counterpart. Upon going from the G2/97 to the G3/99 test set, the mean absolute deviation increases from 0.99 to 1.08 kcal/mol, from 1.15 to 1.21 kcal/mol, and from 1.36 kcal/mol to 1.38 kcal/mol, for the G3S, G3S(MP3), and G3S(MP2) theories, respectively.

The increase in the mean absolute deviation for all three methods is primarily due to large errors in the calculated enthalpies of formation of some of the non-hydrogen species in the expanded G3/99 test set. This is similar to the results for the G3 methods based on the higher-level correction per electron pair. The G3S mean absolute deviation of 3.37 kcal/mol for the 13 non-hydrogen species in the G3-3 subset is more than twice that of 1.60 kcal/mol for the 34 non-hydrogens in the G2/97 set. Similar increases in the mean absolute deviations occur for the G3S(MP3) and G3S(MP2) theories.

The mean absolute deviation of G3S for the G3-3 subset of larger molecules is 1.43 kcal/mol compared to 1.30 kcal/mol for G3. The larger increase for G3S method suggests that the scaling approach based on six parameters may not work as well on molecules outside the parameterization test set as does an approach based on the four-parameter higher-level correction. For example, the deviation of G3S standard enthalpy of formation of P_4, a molecule with unusual bonding, amounts to 10 kcal/mol. If nine instead of six scaling parameters are used in G3S, the mean absolute deviation is 1.29 kcal/mol for the new subset, about the same as G3 theory. This suggests that a larger number of parameters may be needed in the scaling approach to make it as accurate as its HLC-based counterpart, although additional assessments are needed to confirm this suspicion.

5. G3X THEORY

A new family of G3 methods, referred to as G3X (G3 eXtended) has been developed recently [30] to improve the accuracy of the results for some of the larger molecules included in the G3/99 test set. In the assessment of G3 theory [21] on the G3/99 test set, the mean absolute deviation from experiment (1.07 kcal/mol) was slightly larger than the corresponding value of 1.01 kcal/mol found originally for the smaller G2/97 test set. Significantly larger deviations were, however, found for the larger non-hydrogen systems containing second-row atoms (Fig. 3.2).

The larger non-hydrogen systems have deviations (3.24 kcal/mol) almost twice as large as those included in the smaller G2/97 test set (1.68 kcal/mol). In particular, hypervalent molecules such as SO_3, SF_6, and PF_5 have deviations ranging from 5 to 7 kcal/mol. Part of the source of errors in the G3 results for the non-hydrogen species was traced to the MP2/6-31G(d) geometries used in the single-point energy calculations. Use of experimental geometries instead of the MP2/6-31G(d) ones in a small subset of non-hydrogen molecules reduced the deviations in those

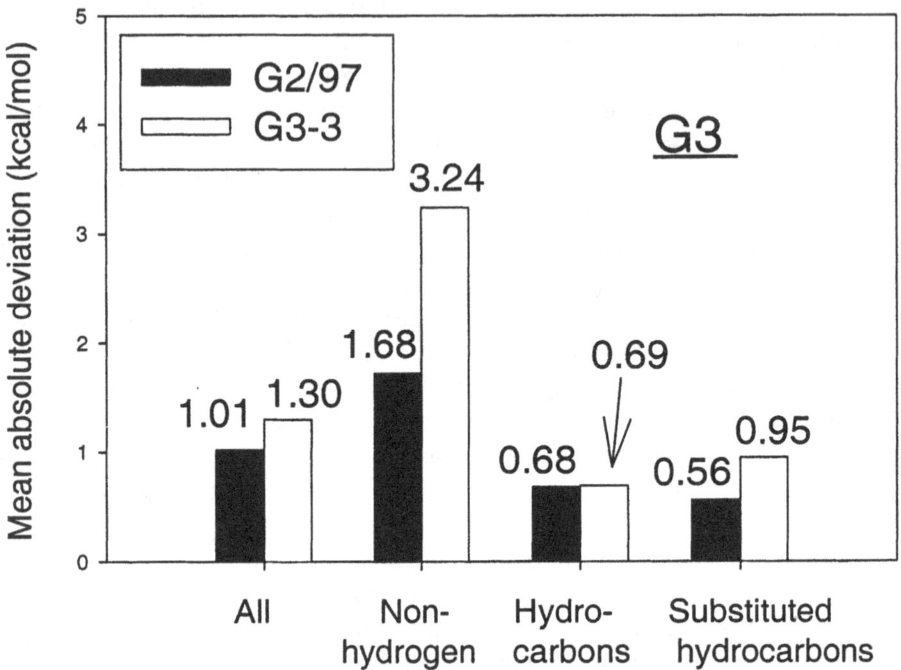

Figure 3.2 Mean absolute deviation from experiment of the G3 predictions for the G3/99 test set broken down into the G2/97 (G2-1 + G2-2) and G3-3 subsets.

species, but they still remained as large as 3 - 4 kcal/mol. The remainder of the error was attributed to basis set deficiencies.

Three modifications have been included in the G3X method to alleviate these deficiencies:

1. B3LYP/6-31G(2df,p) geometries are used in place of the MP2(fu)/ 6-31G(d) geometries. These new geometries have significantly smaller deviations from experiment than their original counterparts. For example, the deviation from experiment in the bond lengths for a subset of seven representative molecules (PF_3, PF_5, P_4, PCl_5, SO_2, SO_3, and SF_6) decreased from 0.027 Å [MP2(fu)/6-31G(d)] to 0.011 Å [B3LYP/6-31G(2df,p)].

2. B3LYP/6-31G(2df,p) zero-point energies (scaled by 0.9854) are used in place of the HF/6-31G(d) zero-point energies (scaled by 0.8929). The former scale factor is derived from fitting the set of zero-point energies compiled by Scott and Radom [57]. The small correction suggests that the B3LYP/6-31G(2df,p) level of theory is

accurate for zero-point energies as well as geometries. This choice differs in two ways from that of the original G3 procedure. First of all, in G3 theory the zero-point energies and geometries are calculated at two different levels of theory, namely HF/6-31G(d) and MP2(fu)/6-31G(d). Secondly, the HF/6-31G(d) scale factor was based on fitting of experimental vibrational frequencies rather than zero-point energies. Thus, the new procedure for calculating zero-point energies in G3X theory is more consistent than that employed in G3 theory.

3. The G3Large basis set for second-row atoms is augmented at the Hartree-Fock level with g valence polarization functions [31]. Significant improvement in the calculated atomization energies were found for some representative molecules. For example, the addition of a single set of g functions to the second-row atoms (Si - Cl) increases the binding in SiF_4, PF_5, and SF_6 by 3.6, 5.1, and 5.5 kcal/mol, respectively, at the Hartree-Fock level. The increase is much smaller for similar molecules containing chlorine. Addition of more polarization functions (2g,2gh) on the second-row atoms results in substantially smaller changes in the atomization energies [31]. Thus, in G3X theory, a single set of g polarization functions (7 pure functions) is added to the second-row G3Large basis set at the HF level. The g exponents for Al - Cl are taken from Dunning's correlation consistent cc-pVQZ basis set [58] (Al: 0.357, Si: 0.461, P: 0.597, S: 0.683, Cl: 0.827, and Ar: 1.007). No g functions are used for Na or Mg. This new basis set is referred to as G3XLarge. It should be noted that similar basis set deficiencies occur also at correlated levels. Correcting such deficiencies at the correlated level is more difficult due to their slow convergence, though the HLC parameters offer partial remedy.

The total G3X energy incorporating these three features is given by the equation

$$\begin{aligned} E^{o}_{\mathrm{G3X}} &= E_{\mathrm{MP4/d}} + (E_{\mathrm{QCISD(T)/d}} - E_{\mathrm{MP4/d}}) \\ &+ (E_{\mathrm{MP4/plus}} - E_{\mathrm{MP4/d}}) + (E_{\mathrm{MP4/2df,p}} - E_{\mathrm{MP4/d}}) \\ &+ (E_{\mathrm{MP2(fu)/G3Large}} - E_{\mathrm{MP2/2df,p}} - E_{\mathrm{MP2/plus}} + E_{\mathrm{MP2/d}}) \\ &+ (E_{\mathrm{HF/G3XLarge}} - E_{\mathrm{HF/G3Large}}) + E_{\mathrm{SO}} + E_{\mathrm{HLC}} + E_{\mathrm{ZPE}} \,. \end{aligned} \tag{5.1}$$

Eq. (5.1) is the same as for G3 theory except for the addition of the Hartree-Fock term. This term extends the HF/G3Large energy, which

Table 3.3 Deviations from experiment for the calculated enthalpies of formation of some non-hydrogen systems (kcal/mol).[a]

Molecule	G3	G3X	G3S	G3SX
SO_2	-3.81	-0.73	-2.00	0.40
SO_3	-5.17	-1.54	-2.15	0.54
PF_3	-4.84	-1.85	-6.21	-2.96
PF_5	-7.05	-1.80	-7.32	-1.18
SF_6	-6.22	-0.47	-4.23	2.66
SiF_4	-1.12	2.27	-1.76	2.48
$SiCl_4$	0.02	-0.63	-1.97	0.13
P_4	-4.15	-2.18	-10.01	-8.79
PCl_3	-3.19	-3.30	-5.04	-3.15
PCl_5	2.40	1.74	1.09	3.44
$POCl_3$	-3.07	-2.32	-3.94	-1.88
SO_2Cl_2	-4.37	-2.55	-2.32	0.11

[a] ΔH_f^0(exp.) – ΔH_f^0(theor.).

is part of the MP2(fu)/G3Large energy, to the G3XLarge basis set. As in G3 theory, all correlation calculations (except for MP2/G3Large) are done with a frozen core.

As discussed earlier, the single-point energies in Eq. (5.1) are calculated at B3LYP/6-31G(2df,p) geometries, and the zero-point energies and thermal corrections are obtained from scaled frequencies computed at the same level of theory. The higher-level correction (HLC) parameters were obtained by fitting to the full G3/99 test set. Fitting of the HLC parameters to the smaller G2/97 test set gives nearly the same values for the four parameters, indicating that there is little sensitivity to the increase in the data set size. The G3X method takes about 10 - 15 % more time than G3 due to the B3LYP/6-31G(2df,p) frequency calculation.

G3X theory gives significantly better agreement with experiment for the G3/99 test set of 376 energies. Overall, the mean absolute deviation from experiment decreases from 1.07 kcal/mol (G3) to 0.95 kcal/mol (G3X). The mean absolute deviation for the 222 enthalpies of formation decreases from 1.05 kcal/mol (G3) to 0.88 kcal/mol (G3X). The improvement is largely due to the non-hydrogen systems for which the mean absolute deviation decreases from 2.11 to 1.49 kcal/mol. The increased accuracy of G3X is due to both the use of a new geometry and the larger Hartree-Fock basis set. The latter is especially important for hypervalent molecules.

Unlike in the case of G3 theory, there is very little degradation in G3X theory for larger molecules. The overall mean absolute deviation from experiment is nearly the same for the larger G3/99 test set (0.95 kcal/mol) as it is for its smaller G2/97 counterpart (0.96 kcal/mol). Similarly, the mean absolute deviation from experiment for enthalpies is nearly the same for the larger G3/99 test set (0.88 kcal/mol) as it is for the smaller G2/97 test set (0.86 kcal/mol). It is also important to note that this result is not dependent on the data set used to obtain the HLC parameters since both sets give essentially the same values.

Some examples of the changes observed upon going from G3 to G3X theory for selected non-hydrogen systems are shown in Table 3.3. Significant improvements are seen in most cases, though errors in the range of 2 or 3 kcal/mol remain in some molecules. The reliability of the G3X method is illustrated by the fact that only one molecule (C_2F_4) has an error of more than 4 kcal/mol (note that the experimental enthalpy of formation of C_2F_4 has been recently called into question).

The three new features of G3X theory can also be easily included in the G3(MP3) and G3(MP2) methods. The resulting theories are referred to as G3X(MP3) and G3X(MP2), respectively. The G3X(MP3) energy is given by

$$\begin{aligned} E^{o}_{G3X(MP3)} &= E_{MP4/d} + (E_{QCISD(T)/d} - E_{MP4/d}) \\ &+ (E_{MP3/2df,p} - E_{MP3/d}) \\ &+ (E_{MP2(fu)/G3Large} - E_{MP2/2df,p}) \\ &+ (E_{HF/G3XLarge} - E_{HF/G3Large}) + E_{SO} + E_{HLC} + E_{ZPE}\,, \end{aligned} \tag{5.2}$$

whereas the G3X(MP2) energy reads

$$\begin{aligned} E^{o}_{G3X(MP2)} &= E_{MP4/d} + (E_{QCISD(T)/d} - E_{MP4/d}) \\ &+ (E_{MP2/G3MP2Large} - E_{MP2/d}) \\ &+ (E_{HF/G3XLarge} - E_{HF/G3MP2Large}) \\ &+ E_{SO} + E_{HLC} + E_{ZPE}\,. \end{aligned} \tag{5.3}$$

Eqs. (5.2) and (5.3) are the same as for G3(MP3) and G3(MP2) theories, except for the addition of the Hartree-Fock (HF) term. As in the case of G3X, this term extends the HF energy to the G3XLarge basis set. Again, the single-point energies in Eqs. (5.2) and (5.3) are calculated at the B3LYP/6-31G(2df,p) geometries and the zero-point energies are

Table 3.4 Summary of mean absolute deviations (kcal/mol) for G3X(MP3) and G3X(MP2) methods applied to the G3/99 test set.

Set of data[a]	G3	G3X	G3X (MP3)	G3X (MP2)
Enthalpies of formation (222)	1.05	0.88	1.07	1.05
Non-hydrogens (47)	2.11	1.49	2.05	1.75
Hydrocarbons (38)	0.69	0.56	0.68	0.76
Subst. hydrocarbons (91)	0.75	0.75	0.76	0.78
Inorganic hydrides (15)	0.87	0.81	1.12	1.01
Radicals (31)	0.87	0.76	0.96	1.17
Ionization energies (88)	1.14	1.07	1.16	1.36
Electron affinities (58)	0.98	0.98	1.29	1.51
Proton affinities (8)	1.34	1.21	1.09	0.79
All (376)[b]	1.07 (1.54)	0.95 (1.35)	1.13 (1.63)	1.19 (1.69)

[a] Number of species in each set given in parentheses.

[b] Root mean square deviations given in parentheses.

computed from scaled B3LYP/6-31G(2df,p) frequencies. The higher-level correction (HLC) parameters were obtained by fitting to the G3/99 test set.

Summaries of G3X(MP3) and G3X(MP2) mean absolute deviations from experiment for the G3/99 test set of 376 energies are given in Table 3.4. The overall mean absolute deviations for G3X(MP3) and G3X(MP2) theories are 1.13 and 1.19 kcal/mol, respectively. These are improvements over G3(MP3) and G3(MP2), which have mean absolute deviations of 1.27 and 1.31 kcal/mol, respectively, for the same set of energies. For enthalpies of formation, the mean absolute deviations decrease from 1.29 to 1.07 kcal/mol [G3X(MP3)] and from 1.22 to 1.05 kcal/mol [G3X(MP2)]. Much of the improvement in enthalpies is due to non-hydrogen molecules, although other types of species also improve slightly or stay the same. The G3X(MP3) and G3X(MP2) methods save considerable computational time and have a reasonable accuracy. The ratio of the computational costs for G3X, G3X(MP3), and G3X(MP2) theories is approximately 5:2:1 for a molecule such as benzene.

The three new features of G3X theory can also be included in the G3S method. The resulting theory is referred to as G3SX and the energy is given by

Table 3.5 Mean absolute deviations (kcal/mol) from experiment for the scaled G3S and G3SX methods applied to the G3/99 test set.

Set of data[a]	G3S	G3SX	G3SX (MP3)	G3SX (MP2)
Enthalpies of formation (222)	1.12	0.88	0.90	1.26
Non-hydrogens (47)	2.09	1.60	1.70	2.23
Hydrocarbons (38)	0.79	0.64	0.66	0.67
Subst. hydrocarbons (91)	0.92	0.72	0.65	1.12
Inorganic hydrides (15)	0.63	0.61	0.65	0.98
Radicals (31)	0.86	0.67	0.88	1.06
Ionization energies (88)	1.09	1.05	1.16	1.38
Electron affinities (58)	0.90	1.02	1.32	1.65
Proton affinities (8)	1.17	1.23	1.29	0.70
All (376)	1.08	0.95	1.04	1.34

[a] Number of species in each set given in parentheses.

$$\begin{aligned} E^{o}_{\mathrm{G3SX}} &= E_{\mathrm{HF/d}} + S_{\mathrm{E234}}\,(E_{\mathrm{2/d}} + E_{\mathrm{3/d}} + E_{\mathrm{4/d}}) + S_{\mathrm{QCI}}\,E_{\Delta\mathrm{QCI/d}} \\ &+ S_{\mathrm{HF'}}\,(E_{\mathrm{HF/G3XLarge}} - E_{\mathrm{HF/d}}) + S_{\mathrm{E2'}}\,(E_{\mathrm{2(fu)/G3Large}} - E_{\mathrm{2/d}}) \\ &+ S_{\mathrm{E3'}}\,[(E_{\mathrm{3/plus}} - E_{\mathrm{3/d}}) + (E_{\mathrm{3/2df,p}} - E_{\mathrm{3/d}})] \\ &+ S_{\mathrm{E4'}}\,[(E_{\mathrm{4/plus}} - E_{\mathrm{4/d}}) + (E_{\mathrm{4/2df,p}} - E_{\mathrm{4/d}})] \\ &+ E_{\mathrm{SO}} + E_{\mathrm{ZPE}}\,. \end{aligned} \tag{5.4}$$

Eq. (5.4) is identical to that of the corresponding G3S method, except for the use of the G3XLarge basis set in the Hartree-Fock term instead of the G3Large basis. Also, the single-point energies are calculated at the B3LYP/6-31G(2df,p) geometries and the zero-point energies are obtained from scaled B3LYP/6-31G(2df,p) frequencies. The scaling parameters were obtained by fitting to the G3/99 test set. G3SX has six parameters, one for the Hartree-Fock energy extension and five for the correlation terms. Note that ideally the parameters should be close to one; however, the $S_{\mathrm{E4'}}$ scale factor is 0.66, which may cause problems in some cases (see below). In a similar manner, the methods based on reduced perturbation orders, namely G3SX(MP3) and G3SX(MP2), are derived by adding the three new features to the G3S(MP3) and G3S(MP2) methods, respectively.

A summary of G3SX mean absolute deviations from experiment for the G3/99 test set of 376 energies is given in Table 3.5. Overall, the mean absolute deviation for G3SX method is 0.95 kcal/mol. This is a substantial improvement over G3S theory, which has a mean absolute deviation of 1.08 for the same set of energies. The mean absolute deviation for enthalpies of formation decreases substantially from 1.12 kcal/mol to 0.88 kcal/mol. The improvement is due to the non-hydrogen species (2.09 to 1.60 kcal/mol) as well as substituted hydrocarbons (0.92 to 0.72 kcal/mol), hydrocarbons (0.79 to 0.64 kcal/mol), and radicals (0.86 to 0.67 kcal/mol). However, consideration of the specific deviations for non-hydrogen molecules indicates that G3SX theory does not do as well for them as G3X theory does. Eight of the 222 enthalpies of formation differ by more than 3 kcal/mol (Na_2, AlF_3, C_2F_4, CH_2CHCl, pyrazine, P_4, PCl_5, and PCl_3). The scaling approach is especially poor for P_4, which has an error of 8.8 kcal/mol. This is probably due to the small scaling factor for the MP4 term. Otherwise, the overall accuracy of G3SX theory, as assessed on the G3/99 test set, is very similar in terms of the mean absolute deviations to that attained by G3X theory, suggesting that both types of parameterizations work equally well.

The G3SX method based on the third-order perturbation theory, G3SX(MP3), is especially noteworthy in that it has a mean absolute deviation of 1.04 kcal/mol for the 376 energies in the G3/99 test set and 0.90 kcal/mol for the 222 enthalpies of formation. In this respect, it is as accurate as G3 theory and much less expensive. All of the G3SX methods have the advantage of being suitable for studies of potential energy surfaces.

6. DENSITY FUNCTIONAL THEORY

Density functional methods provide a cost-effective way of treating the electron correlation effects in larger molecules and are being increasingly used for a variety of problems. New functionals are being developed by many groups and the accuracy of the functionals is also improving steadily. Interestingly, some of the popular functionals such as B3LYP have been parameterized based on their performance for the original G2 test set of molecules. It is clearly of interest to compare the performance of the density functional methods with those of G3 theory and its approximate versions for the same test set of molecules.

Three density functional theories (DFT), namely LDA, BLYP, and B3LYP, are included in this section. The simplest is the local spin density functional LDA (in the SVWN implementation), which uses the Slater exchange functional [59] and the Vosko, Wilk and Nusair [60] correlation functional. The BLYP functional uses the Becke 1988 exchange

Table 3.6 Mean absolute deviations (kcal/mol) for heats of formation for different test sets.[a]

Method	G2/97 (147)	G3-3 (75)	G3/99 (222)
G3	0.92	1.30	1.05
G3S	0.96	1.43	1.12
G3X	0.86	0.89	0.88
B3LYP	3.08	8.21	4.81
BLYP	7.25	13.32	9.30
LDA	91.9	216.5	134.0

[a] Number of species in each set given in parentheses.

functional [61], together with the correlation term of Lee, Yang and Parr [62]. The B3LYP functional uses parameters fitted to the data in the original G2 test set and is given by a linear combination of Hartree-Fock exchange, 1988 Becke exchange, and various correlation parts [59, 60]. Functionals such as B3LYP that contain a portion of exact exchange are generally referred to as hybrid density functionals. The 6-311+G(3df,2p) basis set [47] is used in all of the density functional calculations reported here.

The three DFT methods under assessment have a wide range of mean absolute deviations (4.27 to 85.27 kcal/mol) for the energies in the G3/99 test set. Table 3.6 compares the performance of the DFT methods with the different variations of G3 theory for the subset of the 222 heats of formation.

The ordering of the reliability of the methods is similar to the results for the G2/97 test set seen previously. As expected from its known tendency for substantial overbinding, the local density method (LDA) performs poorest with a mean absolute deviation of 134 kcal/mol. The BLYP functional has a mean absolute deviation of 9.3 kcal/mol, while the B3LYP functional performs the best with a mean absolute deviation of 4.8 kcal/mol. In our previous study on the G2/97 test set that included seven functionals, the B3LYP function also had the lowest mean absolute deviation.

There is a significant increase in deviations of the data obtained with DFT methods for the heats of formation in the new G3-3 subset. The B3LYP and BLYP mean absolute deviations for the G3-3 subset are about two times larger than that in the G2/97 test set (8.21 kcal/mol vs. 3.08 kcal/mol and 13.32 kcal/mol vs. 7.25 kcal/mol, respectively).

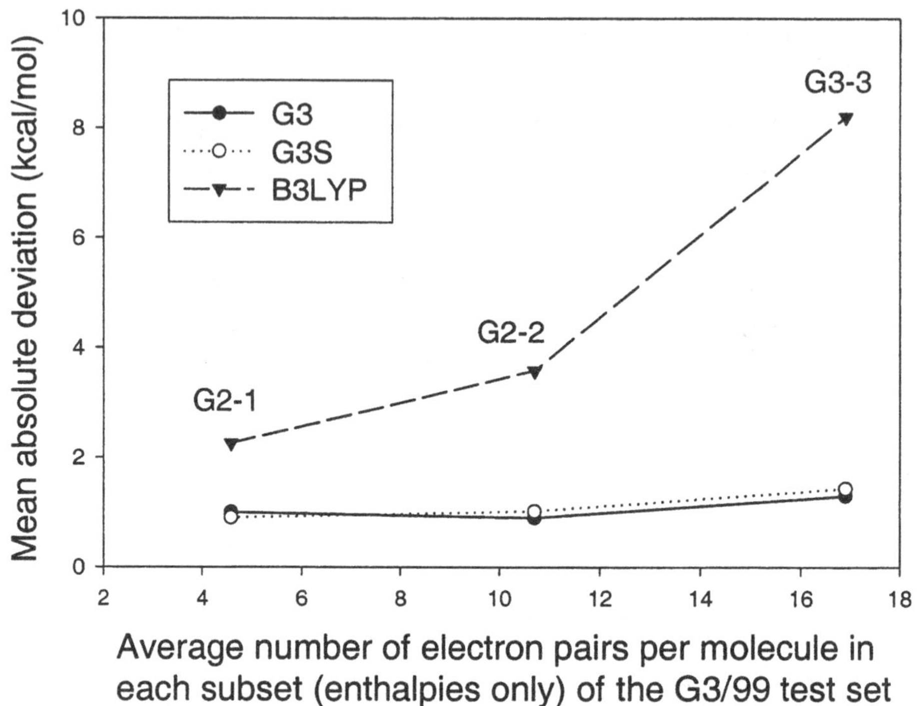

Figure 3.3 Mean absolute deviations for G3, G3S, and B3LYP as a function of the average number of electron pairs in the G2-1, G2-2, and G3-3 subsets (enthalpies of formation only).

The mean absolute deviations of LDA method are much larger (216.5 kcal/mol vs. 91.9 kcal/mol). The increase in the deviations is largest for the hydrocarbons and their substituted derivatives. For B3LYP, the mean absolute deviations increase from 2.92 kcal/mol (G2/97) to 9.64 kcal/mol (G3-3 subset) for the hydrocarbons, and from 2.22 kcal/mol (G2/97) to 7.15 kcal/mol (G3-3 subset) for substituted hydrocarbons. The B3LYP mean absolute deviation for the non-hydrogen species increases from 5.15 kcal/mol to 10.99 kcal/mol. An example of the increase in error with molecular size is evident from comparison of results for propane and n-octane. The B3LYP enthalpy deviates from experiment by -1.46 kcal/mol for propane and -14.04 kcal/mol for n-octane. The G3 deviations for these cases are only 0.33 kcal/mol and 0.88 kcal/mol, respectively. The error per bond in G3 theory is about 0.035 for both propane and n-octane, whereas for B3LYP it is 0.146 and 0.51 kcal/mol, respectively.

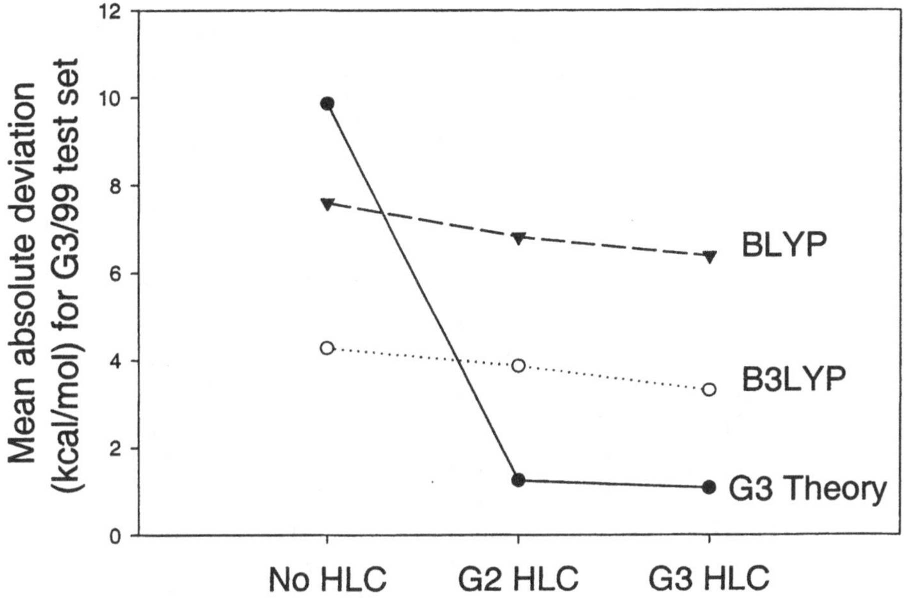

Figure 3.4 Effect of inclusion of higher level corrections on the mean absolute deviations of G3, B3LYP, and BLYP methods for the enthalpies of formation in the G3/99 test set.

The reason for the larger deviations in the new G3-3 subset of energies is that the DFT errors tend to accumulate in the larger molecules [26]. This is evident in Fig. 3.3 that shows a plot of the mean absolute deviation for enthalpies of formation vs. the average number of electron pairs in the species for the three subsets of the G3/99 test set. This is an approximate way to gauge the dependence of the errors on the size of the molecule. The G2-1 subset includes only small molecules (one and two non-hydrogen atoms except for CO_2 and SO_2), while G2-2 includes larger molecules (up to six non-hydrogen atoms), and the new G3-3 subset includes molecules with up to ten non-hydrogen atoms. The plot indicates that the mean absolute deviation increases with the number of pairs of electrons in the subspecies, thus confirming the accumulation of error that occurs in the B3LYP method due to the size of the molecule. Note that this accumulation of error does not occur for ionization energies and electron affinities since only one (or no) electron pair is being broken.

Since application of a higher-level correction for electron pairs works well in the G2 and G3 methods to reduce deficiencies per electron pair,

Table 3.7 Deviations (kcal/mol) in the enthalpies of formation of n-alkanes.[a]

Species	G3	G3(MP3)	G3(MP2)	B3LYP
CH_4	0.25	0.47	-0.05	1.62
C_2H_6	0.31	0.57	0.03	0.60
C_3H_8	0.33	0.58	0.08	-1.46
C_4H_{10}	0.40	0.63	0.18	-3.69
C_5H_{12}	0.35	0.54	0.12	-5.99
C_6H_{14}	0.60	0.75	0.37	-8.07
C_7H_{16}	0.77	0.89	0.54	-10.25
C_8H_{18}	0.88	0.96	0.63	-12.50
C_9H_{20}	(1.29)	(1.33)	1.05	-14.44
$C_{10}H_{22}$	(1.30)	(1.30)	1.06	-16.78
$C_{11}H_{24}$	(1.27)	(1.23)	(1.00)	-19.16
$C_{12}H_{26}$	(1.88)	(1.80)	(1.59)	-20.90
$C_{13}H_{28}$	(1.76)	(1.64)	(1.47)	-23.36
$C_{14}H_{30}$	(1.93)	(1.77)	(1.63)	-25.55
$C_{15}H_{32}$	(1.59)	(1.39)	(1.28)	-28.24
$C_{16}H_{34}$	(1.91)	(1.68)	(1.60)	-30.26

[a] See the text for explanation.

it is of interest to apply it to improve the DFT methods. We have derived higher-level corrections for the B3LYP and BLYP functionals in a manner exactly analogous to those used for the G2 and G3 theories. They were obtained by optimization of the parameters for the G2/97 test set and then applied to the whole G3/99 test set. In Fig. 3.4, the results of applying the G2-like HLC (2 parameters) and G3-like HLC (4 parameters) to B3LYP for the G3/99 test set are shown. Addition of the G2-like HLC improves the mean absolute deviation from 4.27 to 3.87 kcal/mol, while addition of the G3-like HLC reduces the mean absolute deviation to 3.31 kcal/mol. While the latter is a significant improvement, it is much smaller than the improvement for G3 theory shown in Fig. 3.4 (from 9.86 kcal/mol to 1.07 kcal/mol). This suggests that the deficiencies in B3LYP are not systematic and cannot be removed by a simple molecule-independent HLC. Similar results are found for the BLYP method.

A comparison of the performance of G3 theory with DFT methods for larger molecules has been performed for the case of n-alkanes [65]. Table 3.7 shows a comparison of the deviations from experiment in the

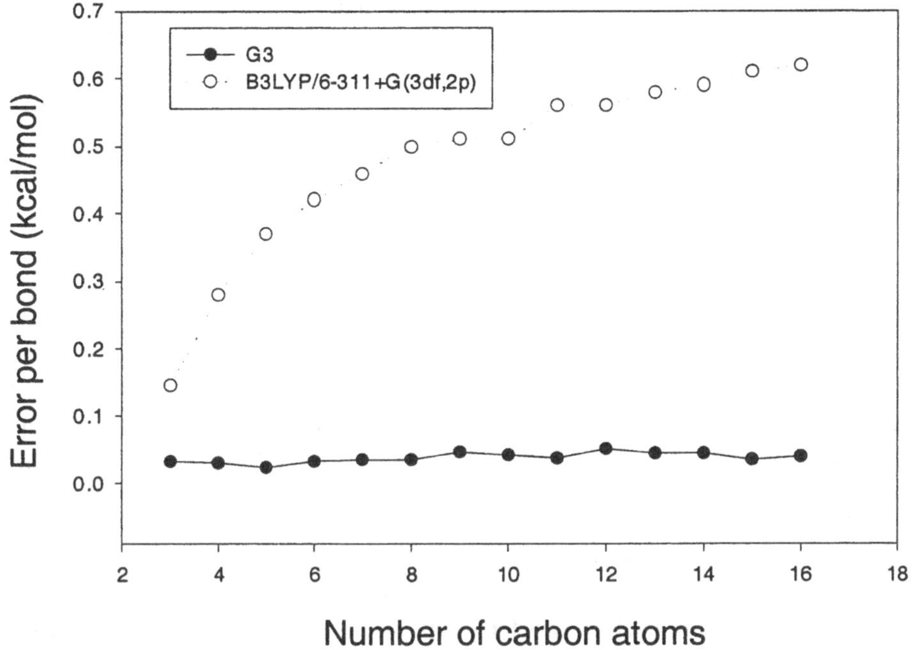

Figure 3.5 Error per bond in the calculated B3LYP enthalpies of formation of n-alkanes.

calculated heats of formation for G3 and B3LYP methods. Values in parentheses denote estimated results for larger n-alkanes based on the corresponding energies of the smaller molecules in the series. The G3, G3(MP3), and G3(MP2) enthalpies of formation of the n-alkanes deviate from experiment by less than 2 kcal/mol. There is evidence of a small accumulation of error (about 0.04 kcal/mol per bond) that increases the overall deviation upon lengthening of the carbon chain. The B3LYP method, however, does very poorly in the calculation of enthalpies of formation for the larger n-alkanes. While the B3LYP enthalpy of formation for propane deviates from experiment by only -1.46 kcal/mol, the deviation for n-hexadecane is -30.3 kcal/mol. This suggests that B3LYP has a significant problem with accumulation of errors in large molecules.

The accumulation of errors in the B3LYP results can be illustrated by the error per bond shown in Fig. 3.5. The error per bond (or electron pair) for the CH_4 to n-$C_{16}H_{34}$ alkanes increases from 0.15 kcal/mol in propane to 0.62 kcal/mol in n-hexadecane, approaching a limit at the longer n-alkanes. From extrapolation to longer n-alkanes, this limit is estimated at about 0.67 kcal/mol. A similar value is obtained by inspecting the CH_2 group energy differences. In contrast, as shown in Fig. 3.5, the error per bond remains constant at about 0.04 kcal/mol for G3 theory. Isodesmic or homodesmotic schemes may be necessary for improving the accuracy of B3LYP results for such large molecules [66, 67].

7. CONCLUDING REMARKS

G3 theory is a general predictive procedure for thermochemical calculations of molecules containing first- and second-row atoms. It has been recently extended to molecules containing third-row non-transition elements. While being computationally more efficient, it constitutes a significant improvement in accuracy over G2 theory. Overall, G3 theory has a mean absolute deviation of 1.07 kcal/mol for the G3/99 test set compared to 1.01 kcal/mol for the G2/97 test set. G3 theory does about as well for the larger hydrocarbons and substituted hydrocarbons in the expanded test set as it does for those in the G2/97 test. However, it does poorly for some of the new and larger non-hydrogen systems in the G3/99 test set such as SF_6 and PF_5, which have errors of 6 - 7 kcal/mol. Part of the source of errors in the G3 results for the non-hydrogen species is traced to the MP2(fu)/6-31G* geometries used in G3 theory. The use of experimental geometries reduces the deviations in those molecules, but they still remain around 3 - 4 kcal/mol. The G3 variants that are based on reduced perturbation orders, G3(MP2) and G3(MP3), perform in a similar manner.

G3 theory based on multiplicative scaling of the energy terms (G3S) instead of the additive higher-level correction has a mean absolute deviation of 1.08 for the G3/99 test set, an increase from 0.99 for the G2/97 test set. As in the case of G3 theory, the increase is largely due to the new non-hydrogen species in the test set. However, systems such as the highly strained P_4 molecule perform poorly with the scaled methods.

G3X theory corrects for most of the shortcomings of G3 theory for larger molecules. It includes better geometries as well as g polarization functions on second-row atoms to correct for the deficiencies of G3 theory for hypervalent molecules. G3X theory gives significantly better agreement with experiment for the G3/99 test set of 376 energies. Overall, the mean absolute deviation from experiment decreases from 1.07 kcal/mol (G3) to 0.95 kcal/mol (G3X). The largest improvement occurs for non-hydrogens for which the mean absolute deviation from experiment decreases from 2.11 to 1.49 kcal/mol. G3X has a mean absolute deviation of 0.88 kcal/mol for the 222 enthalpies of formation in the G3/99 test set. Unlike G3 theory, G3X does not decrease in accuracy for the larger molecules added to the G2/97 test set to form the G3/99 test set. The related G3SX methods have the advantage of being suitable for studies of potential energy surfaces.

The density functional methods assessed in this study (B3LYP, BLYP, and LDA) all perform much worse for the enthalpies of formation of the larger molecules in the G3/99 set. This is due to a cumulative effect in the errors for the larger molecules in this test set. The errors are found to be approximately proportional to the number of pairs of electrons in the molecules but the methods are not improved significantly when a higher-level correction such as that used in G2 or G3 theory is added the DFT methods. Further correction schemes may be necessary to improve the performance of density functional methods for large molecules.

REFERENCES

1. K. K. Irikura and D. J. Frurip (Eds.) *Computational Thermochemistry*, ACS Symp. Series 677, Washington D.C. (1998).
2. S. R. Langhoff (Ed.) *Quantum Mechanical Electronic Structure Calculations with Chemical Accuracy*, Kluwer Academic Publishers, Dordrecht (1995).
3. D. R. Yarkony (Ed.) *Modern Electronic Structure Theory, Parts I and II.* World Scientific, Singapore (1995).
4. C. W. Bauschlicher, Jr. and S. R. Langhoff, *Science* **254**, 394 (1991).
5. R. S. Grev, C. L. Janssen, and H. F. Schaefer, III, *J. Chem. Phys.* **197**, 8389 (1992).
6. J. A. Montgomery, Jr., J. W. Ochterski, and G. A. Petersson, *J. Chem. Phys.* **101**, 5900 (1994).
7. K. A. Peterson and T. H. Dunning, Jr., *J. Chem. Phys.* **106**, 4119 (1997).
8. D. Feller and K. A. Peterson, *J. Chem. Phys.* **108**, 154 (1998).
9. J. M. L. Martin and G. de Oliveira, *J. Chem. Phys.* **111**, 1843 (1999).
10. P. L. Fast, M. L. Sanchez, and D. G. Truhlar, *J. Chem. Phys.* **111**, 2921 (1999).
11. K. Raghavachari, G. W. Trucks, J. A. Pople, and M. Head-Gordon, *Chem. Phys. Lett.* **157**, 479 (1989).
12. J. A. Pople, M. Head-Gordon, and K. Raghavachari, *J. Chem. Phys.* **87**, 5968 (1987).
13. R. Krishnan, M. J. Frisch, and J. A. Pople, *J. Chem. Phys.* **72**, 4244 (1980).
14. For recent reviews see: L. A. Curtiss and K. Raghavachari, in *Computational Thermochemistry*, K. K. Irikura and D. J. Frurip (Eds.), ACS Symposium Series 677, American Chemical Society, Washington D. C. (1998), pp. 176-197; L. A. Curtiss and K. Raghavachari, in *Encyclopedia of Computational Chemistry*, P. v. R. Schleyer (Ed.), John Wiley, New York (1998).

15. J. A. Pople, M. Head-Gordon, D. J. Fox, K. Raghavachari, and L. A. Curtiss, *J. Chem. Phys.* **90**, 5622 (1989).

16. L. A. Curtiss, C. Jones, G. W. Trucks, K. Raghavachari, and J. A. Pople, *J. Chem. Phys.* **93**, 2537 (1990).

17. L. A. Curtiss, K. Raghavachari, G.W. Trucks, and J.A. Pople, *J. Chem. Phys.* **94**, 7221 (1991).

18. L. A. Curtiss, K. Raghavachari, and J.A. Pople, *J. Chem. Phys.* **98**, 1293 (1993).

19. L. A. Curtiss, J. E. Carpenter, K. Raghavachari, and J. A. Pople, *J. Chem. Phys.* **96**, 9030 (1992).

20. L. A. Curtiss, K. Raghavachari, and J. A. Pople, *J. Chem. Phys.* **103**, 4192 (1995).

21. L. A. Curtiss, K. Raghavachari, P. C. Redfern, V. Rassolov, and J. A. Pople, *J. Chem. Phys.* **109**, 7764 (1998).

22. L. A. Curtiss, P. C. Redfern, K. Raghavachari, and J. A. Pople, *Chem. Phys. Lett.* **313**, 600 (1999).

23. L. A. Curtiss, P. C. Redfern, K. Raghavachari, V. Rassolov, and J. A. Pople, *J. Chem. Phys.* **110**, 4703 (1999).

24. L.A. Curtiss, K. Raghavachari, P. C. Redfern, and J.A. Pople, *J. Chem. Phys.* **106**, 1063 (1997).

25. L. A. Curtiss, P. C. Redfern, K. Raghavachari, and J.A. Pople, *J. Chem. Phys.* **109**, 42 (1998).

26. L. A. Curtiss, K. Raghavachari, P. C. Redfern, and J. A. Pople, *J. Chem. Phys.* **112**, 7374 (2000).

27. A. G. Baboul, L. A. Curtiss, P. C. Redfern, and K. Raghavachari, *J. Chem. Phys.* **110**, 7650 (1999).

28. L. A. Curtiss, K. Raghavachari, P. C. Redfern, A. G. Baboul, and J. A. Pople, *Chem. Phys. Lett.* **314**, 101 (1999).

29. L. A. Curtiss, K. Raghavachari, P. C. Redfern, and J. A. Pople, *J. Chem. Phys.* **112**, 1125 (2000).

30. L. A. Curtiss, P. C. Redfern, K. Raghavachari, and J.A. Pople, *J. Chem. Phys.* **114**, 108 (2001).

31. G. A. Petersson in *Computational Thermochemistry*, K. K. Irikura and D. J. Frurip, (Eds.), ACS Symp. Ser. 677 176-196 (1998).

32. M. S. Gordon and D. G. Truhlar, *J. Am. Chem. Soc.* **108**, 5412 (1986).

33. P. E. M. Siegbahn, R. A. M. Blomberg, and M. Svensson, *Chem. Phys. Lett.* **223**, 35 (1994).

34. P. L. Fast, J. C. Corchado, M. L. Sanchez, and D. G. Truhlar, *J. Phys. Chem.* **A 103**, 5129 (1999).

35. P. L. Fast, M. L. Sanchez, J. C. Corchado, and D. G. Truhlar, *J. Chem. Phys.* **110**, 11679 (1999).

36. P. L. Fast, M. L. Sanchez, and D. G. Truhlar, *Chem. Phys. Lett.* **306**, 407 (1999).

37. J. B. Pedley, R. D. Naylor, and S. P. Kirby, *Thermochemical Data of Organic Compounds, Second Edition*, Chapman and Hall, New York (1986).

38. M. W. Chase, Jr., C. A. Davies, J. R. Downey, Jr., D. J. Frurip, R. A. McDonald, A. N. Syverud, *J. Phys. Chem. Ref. Data 14*, Suppl. 1 (1985). JANAF Thermochemical Tables Third Edition.

39. S. G. Lias, J. E. Bartmess, J. F. Liebman, J. L. Holmes, R. D. Levin, and W. G. Mallard, *J. Phys. Chem. Ref. Data 17*, Suppl. 1 (1988). Gas-Phase Ion and Neutral Thermochemistry.

40. R. L. Asher, E. H. Appelman, and B. Ruscic, *J. Chem. Phys.* **105**, 9781 (1996).

41. G. A. Petersson, D. K. Malick, W. G. Wilson, J. W. Ochterski, J. A. Montgomery, Jr., and M. J. Frisch, *J. Chem. Phys.* **109**, 10570 (1999).

42. B. Ruscic, J. V. Michael, P. C. Redfern, L. A. Curtiss, and K. Raghavachari, *J. Phys. Chem.* **A 102**, 10889 (1998).

43. J. M. L. Martin and P. R. Taylor, *J. Phys. Chem.* **103**, 4427 (1999).

44. D. Feller and D. A. Dixon, *J. Phys. Chem.* **103**, 6413 (1999).

45. C. W. Bauschlicher, P. R. Taylor, and J. M. L. Martin, *J. Phys. Chem.* **A 103**, 7715 (1999).

46. L. A. Curtiss, K. Raghavachari, P. C. Redfern, G. S. Kedziora, and J.A. Pople, *J. Phys. Chem.* **105**, 227 (2001).

47. W. J. Hehre, L. Radom, J. A. Pople, P. v. R. Schleyer, *Ab Initio Molecular Orbital Theory*, John Wiley, New York (1987).

48. J. A. Pople, H. B. Schlegel, R. Krishnan, D. J. Defrees, J. S. Binkley, M. J. Frisch, R. A. Whiteside, R. F. Hout, and W. J. Hehre, *Int. J. Quantum Chem. Symp.* **15**, 269 (1981).

49. C. Moore, *Natl. Bur. Stand. (U.S.)*, Circ 467 (1952).

50. J.-P. Blaudeau (private communication). For more information on the methods used to calculate the spin-orbit quantities see J.-P.Blaudeau and L. A. Curtiss, *Int. J. Quant. Chem.* **60**, 943 (1997).

51. G. N. Lewis and M. Randall, *Thermodynamics, 2nd Edition*, revised by K. S. Pitzer and L. Brewer, McGraw-Hill, New York (1961).

52. M. J. Frisch, G. W. Trucks, H. B. Schlegel, G. E. Scuseria, M. A. Robb, J. R. Cheeseman, V. G. Zakrzewski, J. A. Montgomery, Jr., R. E. Stratmann, J. C. Burant, S. Dapprich, J. M. Millam, A. D. Daniels, K. N. Kudin, M. C. Strain, O. Farkas, J. Tomasi, V. Barone, M. Cossi, R. Cammi, B. Mennucci, C. Pomelli, C. Adamo, S. Clifford, J. Ochterski, G. A. Petersson, P. Y. Ayala, Q. Cui, K. Morokuma, D. K. Malick, A. D. Rabuck, K. Raghavachari, J. B. Foresman, J. Cioslowski, J. V. Ortiz, B. B. Stefanov, G. Liu, A. Liashenko, P. Piskorz, I. Komaromi, R. Gomperts, R. L. Martin, D. J. Fox, T. Keith, M. A. Al-Laham, C. Y. Peng, A. Nanayakkara, C. Gonzalez, M. Challacombe, P. M. W. Gill, B. Johnson, W. Chen, M. W. Wong, J. L. Andres, C. Gonzalez, M. Head-Gordon, E. S. Replogle, and J. A. Pople, Gaussian 98, Gaussian, Inc. Pittsburgh, PA, 1998.

53. L. A. Curtiss, P. C. Redfern, V. Rassolov, G. Kedziora, and J. A. Pople, *J. Chem. Phys.*, in press.

54. L. A. Curtiss, M. P. McGrath, J.-P. Blaudeau, N. E. Davis, and R. Binning, *J. Chem. Phys.* **103**, 6104 (1995).

55. J.-P. Blaudeau, M. P. McGrath, L. A. Curtiss, and L. Radom, *J. Chem. Phys.* **107**, 5016 (1997).

56. P. C. Redfern, L. A. Curtiss, and J.-P. Blaudeau, *J. Phys. Chem.* **101**, 8701 (1997).

57. A. P. Scott and L. Radom, *J. Phys. Chem.* **100**, 16502 (1996).

58. K. A. Peterson, D. E. Woon, and T. H. Dunning Jr, *J. Chem. Phys.* **100**, 7410 (1994).

59. J. C. Slater, *The Self-Consistent Field for Molecules and Solids: Quantum Theory of Molecules and Solids* Vol. 4, McGraw-Hill, New York (1974).

60. S. H. Vosko, L. Wilk, and M. Nusair, *Can. J. Phys.* **58**, 1200 (1980).

61. A. D. Becke, *Phys. Rev.* **A 38**, 3098 (1988).

62. C. Lee, W. Yang, and R. G. Parr, *Phys. Rev.* **B 37**, 785 (1988).

63. A. D. Becke, *J. Chem. Phys.* **98**, 5648 (1993).

64. P. J. Stephens, F. J. Devlin, C. F. Chabalowski, and M. J. Frisch, *J. Phys. Chem.* **98**, 11623 (1994).

65. P. C. Redfern, P. Zapol, L. A. Curtiss, and K. Raghavachari, *J. Phys. Chem.* **A 104**, 5850 (2000).

66. K. Raghavachari, B. B. Stefanov, and L. A. Curtiss, *J. Chem. Phys.* **106**, 6764 (1997).

67. K. Raghavachari, B. B. Stefanov, and L. A. Curtiss, *Mol. Phys.* **91**, 555 (1997).

Chapter 4

Complete Basis Set Models for Chemical Reactivity: from the Helium Atom to Enzyme Kinetics

George A. Petersson
Hall-Atwater Laboratories of Chemistry, Wesleyan University, Middletown, Connecticut 06459-0180, U.S.A.

1. INTRODUCTION

The qualitative idea of chemical reactivity is quantitatively expressed in the rate and extent of chemical reactions. These rate and equilibrium constants present a formidable challenge to theoretical predictions [1, 2]. The principal difficulty lies in the extreme sensitivity of the specific rate constant $k_{rate}(T)$ and the equilibrium constant $K_e(T)$ to small errors in the calculated barrier height $\Delta E_0^\ddagger$ and enthalpy change ΔH_T^0. An error of only 1.4 kcal/mol in these energy changes leads to an error of an order of magnitude in $k_{rate}(T)$ or $K_e(T)$ at room temperature. Thus, one needs methods for calculating molecular energy changes with errors less than ca. 0.5 kcal/mol. This is a very demanding standard, requiring convergence of both the one-particle expansion (basis set) and the n-particle expansion (correlation energy) [2]. In the absence of near-degeneracies, the coupled cluster method of Bartlett [3] with Raghavachari's perturbation treatment of triple excitations [4], CCSD(T), is a sufficiently accurate treatment of the n-particle problem [5]. The one-particle expansion is not so easily disposed of.

The slow convergence of the correlation energy with the one-electron basis set expansion has provided the motivation for several attempts to extrapolate to the complete basis set limit [6-13]. Such extrapolations require a well defined sequence of basis sets and a model for the convergence of the resulting sequence of approximations to the

J. Cioslowski (ed.), Quantum-Mechanical Prediction of Thermochemical Data, 99–130.

correlation energy. The various extrapolation schemes that have been proposed differ in both the method used to obtain a well defined sequence of one-electron basis sets and in the extrapolation model. The complete basis set (CBS) extrapolations described in this chapter employ the asymptotic convergence of pair natural orbital (PNO) expansions [8-10, 14].

Early work on atoms [15, 16] employed increasing sets of s, p, d, ... etc. basis functions, explicitly seeking convergence to the complete basis set limit. The power of such methods was greatly enhanced by the classic papers of Schwartz establishing the asymptotic convergence of the second-order Møller-Plessett (MP2) pair correlation energies [17] with the angular momentum expansions [6, 7],

$$
\begin{aligned}
{}^{\alpha\beta}e_{ij}^{(2)}(l \le l_{\max}) &\rightarrow {}^{\alpha\beta}e_{ij}^{(2)}(\mathrm{CBS}) + \frac{15}{256}\,{}^{\alpha\beta}f_{ij}\,(l_{\max} + \frac{1}{2})^{-3} \text{ as } l_{\max} \rightarrow \infty, \\
{}^{\alpha\alpha}e_{ij}^{(2)}(l \le l_{\max}) &\rightarrow {}^{\alpha\alpha}e_{ij}^{(2)}(\mathrm{CBS}) + \frac{15}{256}\,{}^{\alpha\alpha}f_{ij}\,(l_{\max} + \frac{1}{2})^{-5} \text{ as } l_{\max} \rightarrow \infty,
\end{aligned}
\tag{1.1}
$$

where the exponents -3 and -5 apply to opposite spin $\alpha\beta$ and equal spin $\alpha\alpha$ or $\beta\beta$ pairs, respectively. Extrapolations to the complete basis set limit using the asymptotic formulas of Schwartz were employed first by Bunge and later by Jankowski, Malinowski, and Polasik to establish a database of CBS-MP2 limits for closed-shell atoms [18, 19].

Several methods have been developed for establishing the MP2 limit for small molecules. We shall compare three of the most important methods, and a recently proposed combination of two of them that achieves a new level of efficiency in obtaining chemically accurate absolute MP2 energy limits. We conclude with a case study of the extension of these approaches to enzyme kinetics, namely the Δ^5-ketosteroid isomerase-catalyzed conversion of Δ^5-androstene-3,17-dione to the Δ^4 isomer.

2. PAIR NATURAL ORBITAL EXTRAPOLATIONS

Twenty years ago, we extended asymptotic extrapolations to polyatomic molecules by transformation of the Schwartz formulae to a symmetry-independent form based on the total number N of pair natural orbitals (PNOs) [8-10, 14],

$$
{}^{\alpha\beta}e_{ij}^{(2)}(\mathrm{N}) \rightarrow {}^{\alpha\beta}e_{ij}^{(2)}(\mathrm{CBS}) + \frac{25}{512}\,{}^{\alpha\beta}f_{ij}\,(\mathrm{N} + \delta_{ij})^{-1} \text{ as } \mathrm{N} \rightarrow \infty,
$$

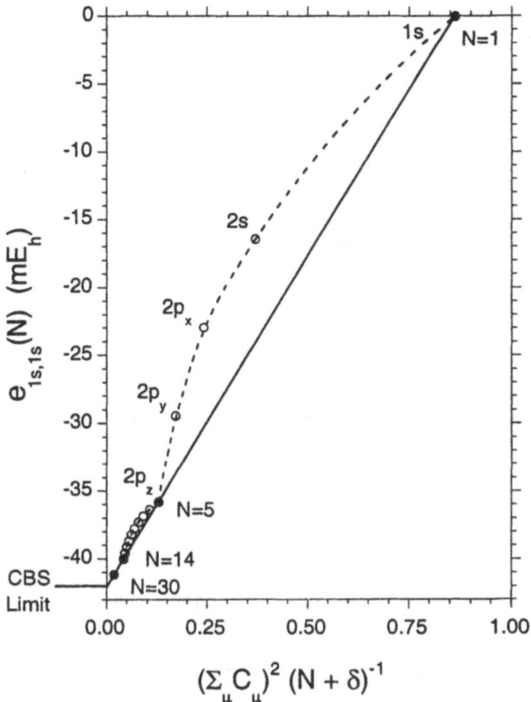

Figure 4.1 Convergence of the correlation energy for the helium ground state with the number N of pair natural orbitals.

$$^{\alpha\alpha}e_{ij}^{(2)}(N) \quad \rightarrow \quad ^{\alpha\alpha}e_{ij}^{(2)}(CBS) + \frac{25}{512}\ ^{\alpha\alpha}f_{ij}\,(N + \delta_{ij})^{-5/3} \text{ as } N \rightarrow \infty, \tag{2.1}$$

where the exponents -1 and -5/3 now apply to opposite spin $\alpha\beta$ and equal spin $\alpha\alpha$ or $\beta\beta$ pairs, respectively. The exclusion parameter δ_{ij} can be determined as the solution of a quadratic equation [20]. Extrapolation of infinite-order (e.g. CCSD) pair energies:

$$e_{ij}^{(\infty)}(N) \rightarrow e_{ij}^{(\infty)}(CBS) + (\sum_{\mu=1}^{N} C_{\mu_{ij}})^2 \left[e_{ij}^{(2)}(N) - e_{ij}^{(2)}(CBS)\right] \text{ as } N \rightarrow \infty, \tag{2.2}$$

requires attenuation by an interference factor $(\Sigma_\mu C_{\mu_{ij}})^2$ obtained from the first-order wavefunction [20].

The algorithm employed for these PNO extrapolations selects the value of N giving the largest (i.e. most negative) value for the CBS pair energy with the constraint of $N \geq N_{min}$. Convergence to the exact CBS

Table 4.1 Comparison of RMS errors[a] in calculated bond lengths (Å, top lines), and harmonic vibrational frequencies (cm^{-1}, bottom lines).

	Level of theory					
Basis Set	HF	B3LYP	MP2	MP3	QCISD	QCISD(T)
3-21G(*)	**0.019**	0.044	0.066	0.042	0.053	0.058
	170	209	279	174	179	188
6-31G*	0.026	0.011	0.019	0.008	0.015	0.019
	136	88	105	83	47	32
6-311G**	0.033	**0.005**	0.009	0.008	**0.005**	0.007
	133	**50**	88	63	**39**	33
6-311+G(2df,2pd)	0.034	0.006	0.009	0.011	0.006	0.004
	134	55	87	63	34	16

[a] Test set: H_2, D_2, N_2, O_2, F_2, CF, CO, CH, NH, OH, and HF.

pair energy is ensured by systematically increasing N_{min} as the basis set is expanded (Fig. 4.1). For example, we generally set N_{min} equal to 5 for spd basis sets and equal to 10 for spdf basis sets. These nonlinear extrapolations are size-consistent only if the canonical SCF orbitals are localized prior to extrapolation of each of the individual pair energies to the CBS limit. We assume that N is large enough for the asymptotic form to be applicable and that the low-lying natural orbitals are accurately described with the basis set employed. Early implementations of these extrapolations served as polyatomic benchmarks for their time [21], but improvements in hardware and software now make more demanding standards possible.

3. CURRENT CBS MODELS

The order-by-order contributions to chemical energies, and thus the number of significant figures required, generally decrease with increasing order of perturbation theory. The general approach for our CBS-n models [20-25] is therefore to first determine the geometry and the zero-point energy (ZPE) at a low level of theory, and then perform a series of high-level single-point electronic energy calculations at this geometry, using large basis sets for the SCF calculation, medium basis sets for the MP2 calculation, and small basis sets for the higher-order calculations through order n. The components of each model have been selected to be balanced so that no single component dominates either the

Table 4.2 The components of the current CBS models.

		CBS-4M	CBS-QB3	CBS-QCI/APNO
Geometry	Method	UHF/3-21G(*)	B3LYP/6-311G(2d,d,p)	QCISD/6-311G(d,p)
ZPE	Method	UHF/3-21G(*)	B3LYP/6-311G(2d,d,p)	QCISD/6-311G(d,p)
	Scale	0.917	0.990	0.989
Core	Method	None	3.92q(Na)+2.83q^2(Na)	$CBS^{(2)}$/KK,KL,LL,L'
SCF	Basis set			
	H, He	31+G(p)	311+G(2p)	APNO 4s2p1d
	Li-Ne	6-311+G(2df)	6-311+G(2df)	6s6p3d2f
	Na-Ar	6-311+G(3d2f)	6-311+G(3d2f)	N/A
MP2	CBS:N_{min}	5	10	10
	Basis set			
	H, He	31G(p')	311+G(2p)	APNO 4s2p1d
	Li-Ne	6-31+G(d')	6-311+G(2df)	6s6p3d2f
	Na-Ar	6-31+G(d')	6-311+G(3d2f)	N/A
MP4	Method	MP4(SDQ)	MP4(SDQ)	
	Basis set			
	H, He	31G	31+G(p')	
	Li-Ne	6-31G	6-31+G(d')	
	Na-Ar	6-31G	6-31+G(d'f')	
CCSD(T)	Method		CCSD(T)	QCISD(T)
	Basis set		6-31+G(d')	6-311+G(2df,p)
Emp	1 e^-	$-5.52\,\Sigma_i \lvert S\rvert^2_{ii} I^2_{ii}$	$-5.79\,\Sigma_i \lvert S\rvert^2_{ii} I^2_{ii}$	$-1.74\,\Sigma_i \lvert S\rvert^2_{ii} I^2_{ii}$
(mE_h)	2 e^-	$-4.55\ (n_\alpha+n_\beta)$		
	Spin	$-38.43\,\Delta\langle S^2\rangle$	$-9.54\,\Delta\langle S^2\rangle$	
G2 error	RMS	2.5	1.1	0.7
(kcal/mol)	Max.	7.0	2.8	1.5

computer time or the error. The CBS-n models use the aforedescribed asymptotic extrapolation to reduce the error from truncation of the basis sets employed in calculation of the correlation energy. The compound model single-point energy is evaluated at a geometry determined at a lower level of theory (e.g. CBS-4M//UHF/3-21G) [24], which is again selected to achieve an accuracy consistent with the single-point energy (Table 4.1). Thus our fastest model, CBS-4M, employs HF/3-21G(*) geometries and frequencies, our intermediate model, CBS-QB3, uses B3LYP/6-311G(2d,d,p) geometries and frequencies, and our most accurate current model, CBS-QCI/APNO, incorporates atomic pair natural orbital (APNO) basis sets and employs QCISD/6-311G** geometries and frequencies [26]. These methods require small empirical corrections to achieve the desired accuracy for chemical energy differences (atomization energies, ionization potentials, and electron affinities) [27]. This sequence of models, namely CBS-4M, CBS-QB3, and CBS-QCI/APNO,

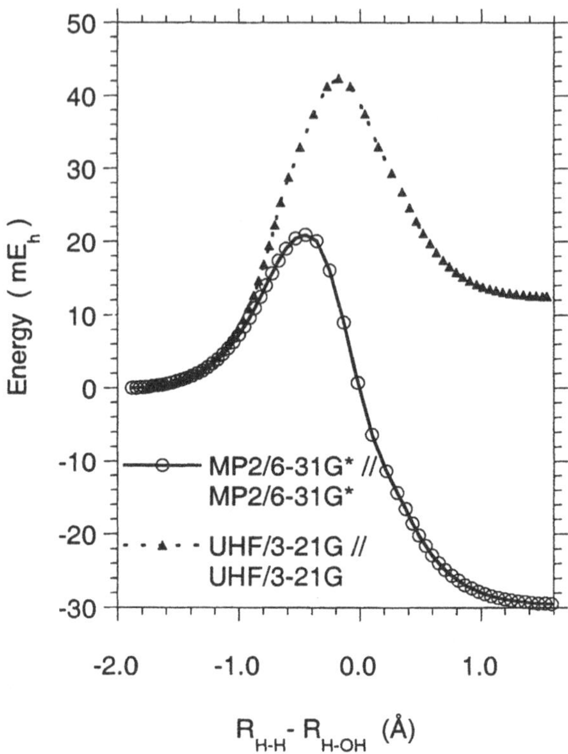

Figure 4.2 The UHF/3-21G energy change along the IRC of the H_2 + •OH → H_2O + •H reaction is qualitatively incorrect.

(Table 4.2), provides a convenient hierarchy, each model reducing both the errors and the maximum size of the molecule accessible by about a factor of two. A user-friendly implementation of these models is readily available within the Gaussian 98^{TM} suite of programs [25].

4. TRANSITION STATES

The development of analytical gradient and Hessian methods [28-36] has made possible the rigorous characterization of transition states within a given level of correlation energy and basis set. The potential energy surface (PES) for a typical bimolecular chemical reaction includes valleys (leading to the reactants and products) connected at the transition state (TS), which is a first-order saddle point (i.e. a stationary point with exactly one negative force constant). The reaction path or intrinsic reaction coordinate (IRC) is defined [37, 38] as the path beginning in

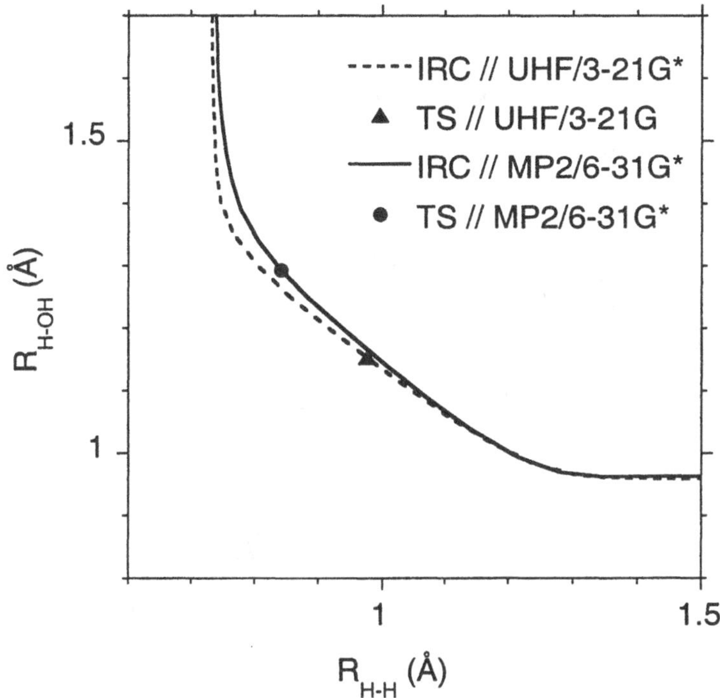

Figure 4.3 The UHF/3-21G IRC of the H_2 + •OH → H_2O + •H reaction is very close to its MP2/6-31G* counterpart.

the direction of negative curvature away from the TS and following the gradient of the PES to the reactants and products.

UHF calculations give notoriously poor results for transition states. For example, the UHF/3-21G energy profile for the transfer of a hydrogen atom from H_2 to •OH is endothermic rather than exothermic and consequently places the transition state too close to the products (Fig. 4.2). One might erroneously conclude that such calculations provide no useful information about the reaction path. Fortunately, this is not the case. Although the variation of the energy along the reaction path is very poorly described by the UHF/3-21G method, the variation of the energy perpendicular to the reaction path is reproduced quite faithfully, just as in stable molecules. Thus, the UHF/3-21G reaction path approximates its MP2/6-31G* counterpart very closely (Fig. 4.3). However, the energy variation along this path, and hence the position of the UHF/3-21G transition state, is incorrect. Nevertheless, the UHF/3-21G reaction path passes through (or near) the MP2/6-31G* transition state. Hence,

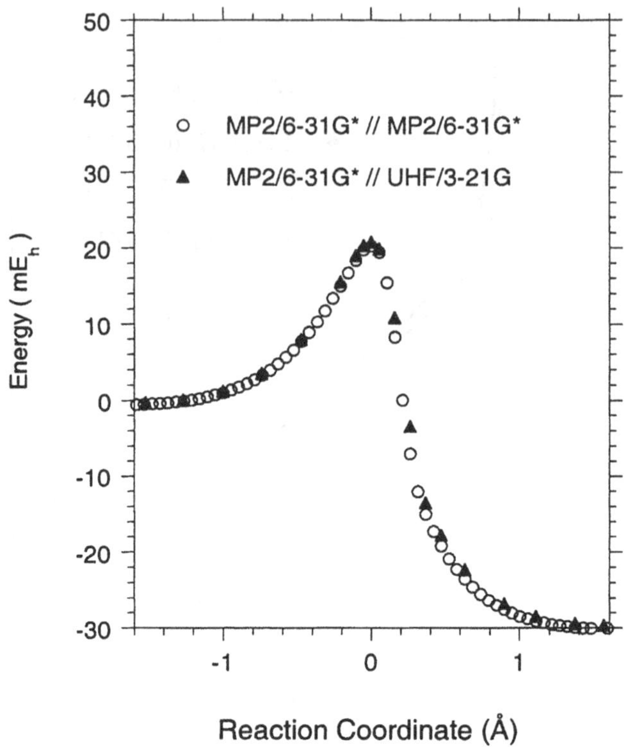

Figure 4.4 The MP2/6-31G* energy along the UHF/3-21G IRC of the H_2 + •OH → H_2O + •H reaction is almost identical to the MP2/6-31G* energy along the corresponding MP2/6-31G* IRC.

if we calculate the MP2/6-31G* energy along the UHF/3-21G reaction path, we obtain an energy profile that differs only marginally from its MP2/6-31G* counterpart along the MP2/6-31G* reaction path (Fig. 4.4). This is really quite remarkable for the H_2 + •OH reaction, given the very poor UHF/3-21G//UHF/3-21G energy profile (Fig. 4.2).

If we move in a direction perpendicular to the reaction path, we find a potential energy curve (or surface) corresponding to a stable reactant or product molecule if we are far from the TS. Even around the TS, the variation of the PES perpendicular to the IRC is very similar to the PES for a stable molecule. Transition states differ from stable molecules in that they possess one negative force constant which defines the reaction coordinate. Calculated energies along the coordinates with positive force constants behave very much like their counterparts in stable molecules. In stark contrast, the energy changes along the reaction coordinate are much more difficult to predict. It is the variation of the energy along this coordinate that is very sensitive to (and thus requires the accurate inclusion of) the correlation energy.

On the basis of the above observations, we have developed [39, 40] the IRCMax{[Method(1)]:[Method(2)]} transition state method, in which we select the maximum of the high-level Method(1) (MP2/6-31G* in Figs. 4.2 - 4.4) along the low-level IRC obtained from Geom [Method(2)] (UHF/3-21G in our example) calculations. The IRCMax transition state extension of the CBS-n models takes advantage of the enormous improvement (from one to two orders of magnitude) in computational speed [41] that is achieved by using the low-level Geom[Method(2)] (e.g. UHF/3-21G in the case of CBS-4 or CBS-q) IRC calculations. We then perform several single point higher-level Energy[Method(1)] (as described in Table 4.2) calculations along the Geom[Method(2)] reaction path to locate the Energy[Method(1)] transition state, i.e. the maximum of Energy[Method(1)] along the Geom[Method(2)] IRC. Calculations at three points bracketing the transition state are sufficient to permit a parabolic fit to determine the transition state and the activation energy.

Since we determine the maximum of Energy[Method(1)] along a path from reactants to products, the IRCMax method gives a rigorous upper bound to the high-level Method(1) transition state energy. In addition, when applied to the CBS-n models, the IRCMax method reduces to the normal treatment of bimolecular reactants and products, Energy[Method(1)]//Geom[Method(2)]. Thus the IRCMax method can be viewed as an extension of these composite models to transition states.

Changes in ZPEs along the reaction path can either increase (by 0.6 kcal/mol for H_2 + •OH) or decrease (by 0.4 kcal/mol for H_2 + •F) the barrier height, depending on the stiffness of the bending force constant. Accurate rate constants can only be obtained if we include the ZPE in our determination of the IRCMax transition state geometry and energy. Our TS algorithm is thus an adaptation of Truhlar's zero-curvature variational transition state theory (ZC-VTST) [42, 43] to our CBS-n models through the use of the IRCMax technique [2, 40].

We selected five hydrogen abstraction reactions for the initial test of our methodology (Fig. 4.5). The barrier heights for these reactions range from 1.3 kcal/mol (H_2 + •F) to 20.6 kcal/mol (H_2O + •H). We included temperatures from 250 K to 2500 K. The rate constants range from 10^{-18} to 10^{-10} cm^3 $molecule^{-1}$ sec^{-1}. All absolute rate constants yielded by our IRCMax{CBS-QCI/APNO:QCISD/6-311G(d,p)} model are within the uncertainty of the experiments [44] (Fig. 4.5). The dashed curves and open symbols for H_2 + •OH, H_2 + •D, and D_2 + •H represent the least-squares fits of smooth curves to large experimental data sets

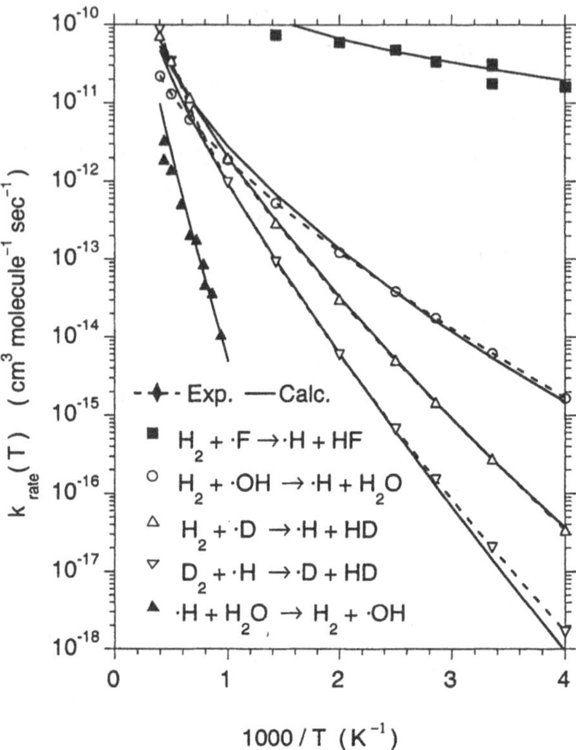

Figure 4.5 Comparison of ZC-VTST IRCMax{CBS-QCI/APNO} absolute rate constants with experiment.

in an attempt to reduce the noise level [44]. The close agreement with theory suggests that this attempt was successful.

Having verified the accuracy of the composite IRCMax{CBS-QCI/APNO:QCISD/6-311G(d,p)} method through comparison with experiment, we then employed these calculations as benchmarks to determine the accuracy of the less demanding CBS-4M and CBS-QB3 models. We employed six reactions [40],

$$H_2 + \bullet CN \rightarrow \bullet H + HCN\,, \tag{4.1}$$

$$CH_4 + \bullet OH \rightarrow \bullet CH_3 + H_2O\,, \tag{4.2}$$

$$\bullet F + HF \rightarrow FH + \bullet F\,, \tag{4.3}$$

$$CH_4 + \bullet CH_3 \rightarrow \bullet CH_3 + CH_4\,, \tag{4.4}$$

$$NCH + \bullet CN \rightarrow \bullet NC + HCN \tag{4.5}$$

and

$$HF + \bullet H \rightarrow \bullet H + FH\,, \tag{4.6}$$

Table 4.3 The RMS error for the calculated TS geometries and barrier heights.

Method	Geometry (Å)	Barrier height (kcal/mol)	ΔH_r^{298} (kcal/mol)
UHF/3-21G	0.22	14.8	25.4
MP2/6-31G*	0.10	7.1	16.3
QCISD/6-311G**	0.09	4.4	5.3
IRCMax CBS-4	0.07	1.3	1.4
CBS-Q	0.03	0.6	0.8

in addition to those displayed in Fig. 4.5. The enthalpies of reaction ΔH_r^{298} were compared with experiment [40], while the TS geometries and the barrier heights were compared with the IRCMax{CBS-QCI/APNO: QCISD/6-311G(d,p)} values (Table 4.3). The barrier heights calculated with the CBS-4 and CBS-Q models are comparable in accuracy to the enthalpies of reaction obtained with these methods if we use the IRCMax extension for the transition states [2, 40].

5. EXPLICIT FUNCTIONS OF THE INTERELECTRON DISTANCE

The singularity in the interelectron Coulomb potential r_{ij}^{-1} creates a cusp in the exact solution of the Schrödinger equation,

$$\Psi(r_{ij}) \rightarrow \Psi(0)\left(1 + \frac{1}{2}r_{ij} + \ldots\right) \quad \text{as } r_{ij} \rightarrow 0\,. \tag{5.1}$$

This is the reason for the expansions in one-electron basis sets being so slowly convergent [8]. An ingenious method for explicitly including this electron coalescence cusp through the resolution of the identity has been developed by Kutzelnigg and Klopper [45]. The details are given in several recent reviews [46, 47], and a recent comparison with one-electron basis set methods is particularly recommended [48]. The cusp is explicitly built into these wavefunctions, but large one-electron basis sets are still required both to accurately describe the remainder of the wavefunction and to converge the resolution of the identity. Thus, Klopper et al. employ [13s8p6d5f/7s5p4d] one-electron basis sets to determine both the MP2 and CCSD(T) limits [48].

The development of these explicit-r_{ij} methods has yielded a database of benchmark results for small polyatomic molecules. These calculations are listed as MP2-R12 and CCSD(T)-R12 in our tables. We have selected the version called MP2-R12/A as a benchmark reference for our study of the convergence to the MP2 limit. This is the version that Klopper et al. found to agree best with our interference effect. The close agreement with extrapolations of one-electron basis set expansions justifies this choice.

6. THE cc-pVnZ BASIS SETS

The Dunning sequences of correlation-consistent basis sets [12, 49] provide a well defined sequence of convergent approximations through the systematic construction of basis sets rather than the projection of pair natural orbitals after completion of the MP2 calculation. Atomic pair natural orbitals (APNOs) form shells, each member of which makes a similar contribution to the correlation energy [8]. Linear combinations of these APNOs produce the corresponding molecular pair natural orbitals [21], making the APNOs a sensible choice for calculations of molecular correlation energies. Adding each new shell of APNOs forms a new member of a consistent sequence of basis sets for electron correlation. Dunning has provided just such a systematic sequence of "correlation-consistent" basis sets ranging from the simple [3s2p1d/2s1p] cc-pVDZ valence double-zeta plus polarization basis sets to the very large [7s6p5d4f3g2h1i/6s5p4d3f2g1h] cc-pV6Z basis sets [12, 49]. Each successive member of the sequence is fully optimized for the neutral atom, and includes one more function of each angular momentum type present in the previous member plus one higher angular momentum function.

The RMS error in the MP2 second-order energy (relative to the MP2-R12 limit determined by Klopper et al. [48]) obtained with the Dunning cc-pVnZ (n = 2 - 6) basis sets for a test set of 12 closed-shell molecules [50] is displayed in Fig. 4.6. This RMS error is reduced by about a factor of two with each increment in the size of the basis set, but even with the largest cc-pV6Z basis sets the RMS error is still an unacceptable 8.4 mE_h. This clearly demonstrates the slow convergence of the correlation energy with the basis set size, but ignores the reason for which Dunning developed these systematic sequences of basis sets, i.e. to facilitate well defined extrapolations to the complete basis set limit [12].

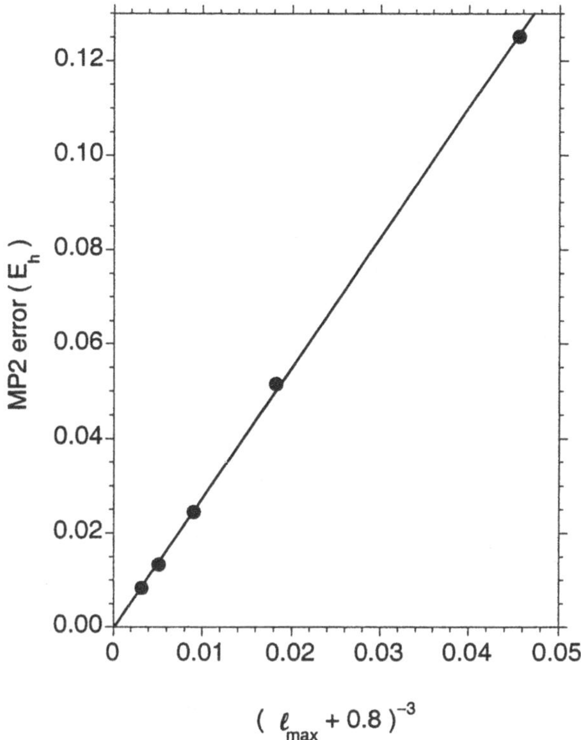

Figure 4.6 The RMS errors in MP2 correlation energies obtained with the cc-pVnZ basis set (n = l_{max} = 2, 3, 4, 5, and 6).

A variety of extrapolation algorithms have been applied to the sequences generated by the correlation-consistent cc-pVnZ basis sets [12, 51-55]. Dunning and his colleagues had initially suggested fitting their calculations to an exponentially decaying function [12, 51, 52],

$$E(n) = E(\infty) + A\, e^{-an} . \tag{6.1}$$

However, as definitive values for $E^{(2)}(CBS)$ became available from the MP2-R12 calculations of Klopper [48], it became clear that Eq. (6.1) seriously underestimates the magnitude of the basis set truncation error. Wilson and Dunning therefore examined [53] a wide variety of extrapolations (24 variations) based on generalizations of Eq. (1.1). They obtained RMS deviations from Klopper's results of less than 1 mE_h using several different extrapolation schemes. We arrived at comparable results (Table 4.4) using just two points, $E^{(2)}(l_{max2})$ and $E^{(2)}(l_{max1})$, so

Table 4.4 Convergence of the $(l_{\max} + \frac{1}{2})^{-3}$ extrapolated cc-pVnZ correlation-consistent basis set MP2 correlation energies (E_h) to the MP2-R12 limit; see Eq. (6.2).

	$E^{(2)}$(DZ,TZ)	$E^{(2)}$(TZ,QZ)	$E^{(2)}$(TZ,5Z)	$E^{(2)}$(TZ,6Z)	MP2-R12
C_2H_2	-0.34128	-0.34664	-0.34625	-0.34600	-0.3465
CH_4	-0.21957	-0.22056	-0.21993	-0.21965	-0.2193
CO	-0.39475	-0.40445	-0.40485	-0.40466	-0.4053
CO_2	-0.66999	-0.68724	-0.68781	-0.68738	-0.6887
H_2	-0.03472	-0.03439	-0.03434	-0.03431	-0.0343
H_2O	-0.29586	-0.30184	-0.30202	-0.30153	-0.3011
HCN	-0.38076	-0.38782	-0.38778	-0.38754	-0.3880
HF	-0.31195	-0.32035	-0.32095	-0.32056	-0.3197
NH_3	-0.26316	-0.26623	-0.26594	-0.26547	-0.2650
N_2	-0.41299	-0.42170	-0.42209	-0.42193	-0.4225
H_2CO	-0.44083	-0.44946	-0.44953	-0.44919	-0.4495
F_2	-0.59631	-0.61150	-0.61297	-0.61269	-0.6136
RMS error	0.00956	0.00100	0.00067	0.00065	

that our single term extrapolation is linear [50],

$$\begin{aligned} E^{(2)}(\infty) &= E^{(2)}(l_{\max 2}) + \left[E^{(2)}(l_{\max 2}) - E^{(2)}(l_{\max 1})\right] \\ &\times \left[\frac{(l_{\max 2} + \frac{1}{2})^{-3}}{(l_{\max 1} + \frac{1}{2})^{-3} - (l_{\max 2} + \frac{1}{2})^{-3}}\right], \end{aligned} \tag{6.2}$$

and thus rigorously size-consistent. In addition, Eq. (6.2) provides a basis for easily obtaining analytical derivatives of the extrapolated MP2 CBS energies.

7. NEW DEVELOPMENTS

The Dunning cc-pVnZ basis sets can be used with our PNO extrapolations to form a potent new combination. We shall consider the SCF energy first, then the MP2 correlation energy, and finally higher-order correlation energy through CCSD(T).

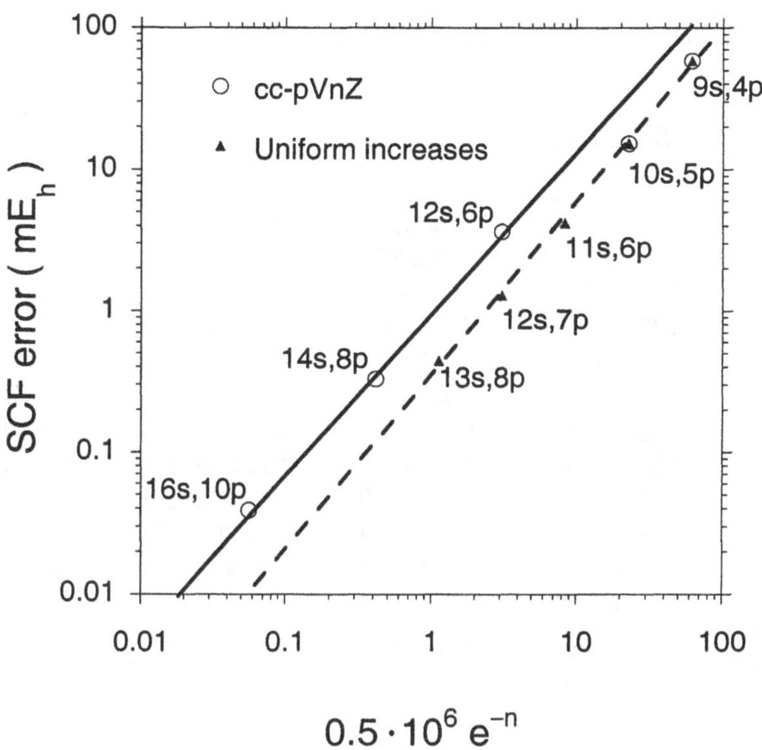

Figure 4.7 The SCF energy of the neon atom converges exponentially with the number of Gaussian primitive basis functions.

7.1. The SCF Limit

One might expect a three-point extrapolation based on Eq. (6.1) to apply to the SCF energy, since the exponential convergence of the SCF energy with the number of Gaussian primitives is well documented [54]. Unfortunately, although the number of contracted basis functions increases in a completely smooth and systematic fashion, the number of primitive Gaussian functions does not increase in a uniform pattern for the cc-pVnZ basis sets (Fig. 4.7). The DZ and TZ basis sets include (9s, 4p) and (10s, 5p) sets of Gaussian primitives, respectively. The QZ basis set then increments the s primitives by 2 but continues the pattern of single increments to the p primitives, yielding (12s, 6p). The 5Z and 6Z basis sets increment both the s and the p by 2, providing (14s, 8p) and (16s, 10p) sets of primitives. Thus, Eq. (6.1) applies to the (QZ, 5Z, 6Z) sequence of basis sets (for first-row atoms, the second-row pattern is different), but not to the (DZ, TZ, QZ) or the (TZ, QZ, 5Z) sequence.

Table 4.5 The extrapolated cc-pVnZ SCF energies (E_h).

	SCF(DZ)	SCF(TZ)	SCF(TZ) +0.321(TZ-DZ)	SCF limit
C_2H_2	-76.825557	-76.848978	-76.856496	-76.855254
CH_4	-40.198689	-40.213377	-40.218092	-40.217041
CO	-112.748970	-112.779950	-112.789895	-112.790441
CO_2	-187.650611	-187.706665	-187.724658	-187.724745
H_2	-1.128720	-1.132950	-1.134308	-1.133616
H_2O	-76.026768	-76.057098	-76.066834	-76.067372
HCN	-92.882909	-92.907552	-92.915462	-92.915338
HF	-100.019441	-100.058048	-100.070441	-100.070811
NH_3	-56.195664	-56.217800	-56.224906	-56.224648
N_2	-108.953751	-108.982944	-108.992315	-108.992596
H_2CO	-113.876136	-113.911596	-113.922978	-113.923113
F_2	-198.685577	-198.751914	-198.773208	-198.773224
RMS error	0.046579	0.011351	0.000580	

A linear extrapolation circumvents this problem. If we assume that the exponent a in Eq. (6.1) is universal, we need only two consecutive points to extrapolate,

$$E(\infty) = E(n+1) + (e^{a} - 1)^{-1} [E(n+1) - E(n)] , \qquad (7.1)$$

which would make extrapolations based on the relatively inexpensive DZ and TZ calculations possible. Empirically, we find that setting a equal to $\sqrt{2}$ in Eq. (7.1) and using n = 2 (i.e. employing the cc-pVDZ and cc-pVTZ SCF energies to extrapolate to the SCF limit) reduces the error in the cc-pVTZ SCF energies by more than an order of magnitude (Table 4.5). Thus, SCF energies with absolute errors of less than 0.5 kcal/mol are available from relatively inexpensive calculations.

7.2. The CBS Limit for the MP2 Correlation Energy

We have recently employed the Dunning correlation-consistent basis sets for our pair natural orbital CBS extrapolation algorithm, Eqs. (2.1) and (2.2) [50]. The results produced a substantial improvement over the raw second-order energies, but were inferior to the $(l_{\max} + \frac{1}{2})^{-3}$ extrapolations listed in Table 4.4. The residual underestimation of

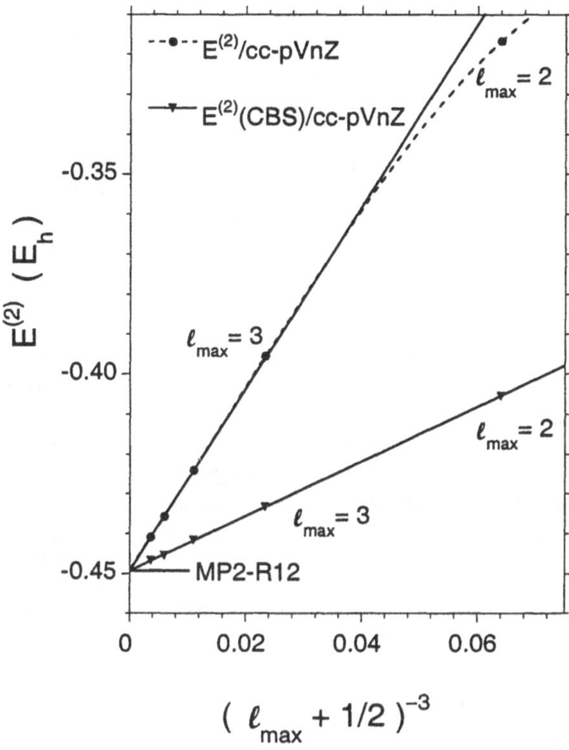

Figure 4.8 The MP2 correlation energy of formaldehyde H_2CO calculated with the Dunning cc-pVnZ basis set (bullets) converges smoothly with $(l_{max} + \frac{1}{2})^{-3}$, where l_{max} is the maximum angular momentum for the basis set. The function becomes linear for $l_{max} \geq 3$. The CBS PNO extrapolated energies (triangles) show a linear convergence with $(l_{max} + \frac{1}{2})^{-3}$, beginning with $l_{max} = 2$, and thus permit such extrapolations from smaller basis sets. These sequences are both in good agreement with the MP2-R12 results of Klopper et al. [48].

the magnitude of the second-order energy component after pair natural orbital extrapolations has two possible origins. Either the number of PNOs employed for the extrapolation was too small for the asymptotic formulae in Eqs. (2.1) and (2.2) to be applicable, or the correlation-consistent basis sets did not describe these PNOs to a sufficient accuracy. The former was less likely since the relative performance of the PNO extrapolations did not improve with increasing l_{max}. In either case, we expected (and actually found) a correlation of this residual error with $(l_{max}+\frac{1}{2})^{-3}$, as indicated in Fig. 4.8 for the example of the formaldehyde molecule. One extrapolation was good, but two were better. The agreement in Fig. 4.8 between the two types of extrapolation and the MP2-R12 limit is striking.

Table 4.6 Convergence of the $(l_{\max}+\frac{1}{2})^{-3}$ extrapolated complete basis set second-order CBS2/cc-pVnZ correlation-consistent basis set MP2 correlation energies (E_h) to the MP2-R12 limit; see Eq. (2.1), (2.2), and (6.2).

	$E^{(2)}$(DZ,TZ)	$E^{(2)}$(DZ,QZ)	$E^{(2)}$(DZ,5Z)	$E^{(2)}$(DZ,6Z)	MP2-R12
C_2H_2	-0.34749	-0.34673	-0.34650	-0.34606	-0.3465
CH_4	-0.21976	-0.21978	-0.21974	-0.21942	-0.2193
CO	-0.40377	-0.40414	-0.40502	-0.40480	-0.4053
CO_2	-0.68599	-0.68671	-0.68794	-0.68750	-0.6887
H_2	-0.03457	-0.03443	-0.03435	-0.03431	-0.0343
H_2O	-0.30185	-0.30144	-0.30180	-0.30145	-0.3011
HCN	-0.38888	-0.38803	-0.38818	-0.38771	-0.3880
HF	-0.31970	-0.32047	-0.32097	-0.32070	-0.3197
NH_3	-0.26544	-0.26547	-0.26566	-0.26521	-0.2650
N_2	-0.42275	-0.42180	-0.42245	-0.42215	-0.4225
H_2CO	-0.44923	-0.44905	-0.44949	-0.44917	-0.4495
F_2	-0.61043	-0.61310	-0.61368	-0.61337	-0.6136
RMS error	0.00137	0.00079	0.00053	0.00053	

The numerical results of this double extrapolation are presented in Table 4.6. The improvement is dramatic for the $(l_{\max}+\frac{1}{2})^{-3}$ extrapolation of the cc-pVDZ and cc-pVTZ PNO extrapolated results, giving an *absolute* accuracy of *better than 1 kcal/mol* with the largest calculation again using just a [4s3p2df/3s2pd] basis set. These calculations are quite routine for molecules as large as naphthalene! Application to several C_{20} species required one to two days each (depending on the specific example) on an SGI Origin 2000 with 8 193 MHz R10000 processors running Gaussian 98 [50].

The PNO extrapolations in Fig. 4.8 and Table 4.6 require localization of the occupied SCF orbitals to ensure size-consistency. In order to preserve this size-consistency for the CBS PNO extrapolations, we have restricted these $(l_{\max}+\frac{1}{2})^{-3}$ extrapolations to a linear form, Eq. (6.2). The new double extrapolation employs this linear extrapolation of pairs of CBS2/cc-pVnZ calculations and thus is rigorously size-consistent. Note that the nonlinear N-parameter $(l_{\max}+\mathrm{a})^{-\alpha}$ extrapolations using least-squares fits to more than N cc-pVnZ energies are *not* size-consistent [53, 55].

Without any extrapolation, energies computed with even the very large [7s6p5d4f3g2h1i/6s5p4d3f2g1h] cc-pV6Z basis sets are still 5.3 kcal/mol from the MP2-R12 limit for our test set of 12 small molecules.

Table 4.7 Convergence of the scaled PNO extrapolated CBS/cc-pVnZ, correlation-consistent basis set higher-order [i.e. CCSD(T)-MP2] correlation energies (E_h) to the CCSD(T)-R12 limit.

	cc-pVDZ	cc-pVTZ	cc-pVQZ	cc-pV5Z	cc-pV6Z	CCSD(T) -R12
C_2H_2	-0.0218	-0.0212	-0.0204	-0.0204	-0.0203	-0.0201
CH_4	-0.0229	-0.0222	-0.0216	-0.0213	-0.0213	-0.0213
CO	-0.0122	-0.0124	-0.0123	-0.0122	-0.0120	-0.0120
CO_2	-0.0052	-0.0064	-0.0070	-0.0071	-0.0070	-0.0069
H_2	-0.0075	-0.0069	-0.0067	-0.0066	-0.0066	-0.0068
H_2O	-0.0084	-0.0081	-0.0080	-0.0079	-0.0077	-0.0074
HCN	-0.0146	-0.0141	-0.0137	-0.0136	-0.0136	-0.0133
HF	-0.0027	-0.0027	-0.0032	-0.0032	-0.0031	-0.0034
NH_3	-0.0157	-0.0152	-0.0147	-0.0145	-0.0145	-0.0140
N_2	-0.0085	-0.0088	-0.0085	-0.0085	-0.0084	-0.0087
H_2CO	-0.0189	-0.0183	-0.0181	-0.0180	-0.0178	-0.0178
F_2	-0.0110	-0.0118	-0.0130	-0.0135	-0.0133	-0.0146
Scaling factor	0.31	0.68	0.96	0.99	0.95	
RMS error	0.0015	0.0011	0.0006	0.0004	0.0004	

In contrast, a linear size-consistent $(l_{max} + \frac{1}{2})^{-3}$ extrapolation of just the MP2/cc-pVTZ and MP2/cc-pVQZ energies is accurate to ±0.63 kcal/mol (Table 4.4). If we try to further reduce the basis sets to cc-pVDZ and cc-pVTZ, the error in the extrapolation increases to ±6.0 kcal/mol. However, the new double extrapolation provides the complete basis set MP2 limit with an absolute accuracy of ±0.86 kcal/mol without recourse to basis sets larger than cc-pVTZ [4s3p2d1f/3s2p1d] (Table 4.6).

7.3. The Higher-Order Correlation Energy

The higher-order contributions to the correlation energy [such as CCSD(T)-MP2] are more than an order of magnitude smaller than their second-order counterparts. However, the basis set convergence to the CCSD(T)-R12 limit does not follow the simple linear behavior found for the second-order correlation energy. This is a consequence of the interference effect described in Eq. (2.2). The full CI or CCSD(T) basis set truncation error is attenuated by the interference factor (Fig. 4.9). The CBS correction to the higher-order components of the correlation energy is thus the difference between the left-hand sides of Eqs. (2.2) and

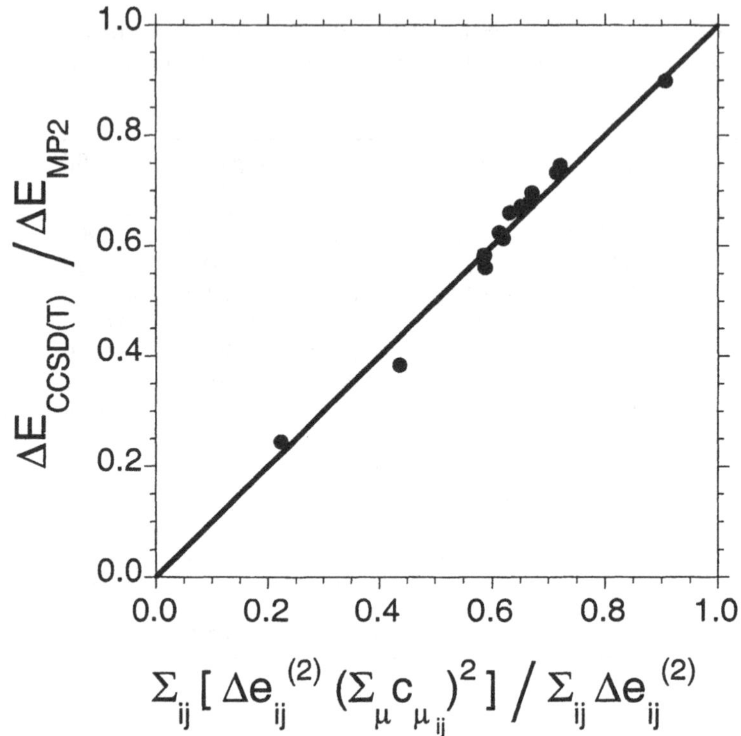

Figure 4.9 The interference effect in Eq. (2.2) gives a quantitative description of the relationship between the MP2 and CCSD(T) basis set truncation errors. These calculations used [5s4p3d2f/4s3p2d] basis sets for the species: Be, H_2, C_2H_2, CH_4, HCN, NH_3, N_2, H_2CO, CO, H_2O, CO_2, HF, F_2, and O^{6+}.

(2.1). This CBS extrapolation reduces the errors in the cc-pVQZ and cc-pV5Z higher-order correlation energy by an order of magnitude (Table 4.7), but seriously over-corrects the cc-pVDZ and cc-pVTZ higher-order energies [50]. A simple scaling to reduce the CBS correction to the cc-pVDZ and cc-pVTZ energies reduces the RMS errors below 1 kcal/mol for both (Table 4.7). This single adjustable CBS higher-order parameter might be compared to the use of a single adjustable parameter in the W1 theory [55].

7.4. Total Energies

Having established that size-consistent extrapolations of energies obtained with the cc-pVDZ and cc-pVTZ basis sets are capable of producing sub-kcal/mol absolute accuracy for SCF energies (Table 4.5),

Table 4.8 Calculated reaction energies (kcal/mol) obtained from the proposed CBS/cc-pVTZ method compared to $(l_{max} + \frac{1}{2})^{-3}$ extrapolation of the CCSD(T)/cc-pVDZ, TZ, QZ, 5Z, and 6Z energies.

Reaction	New CBS	CCSD(T)/DZ...6Z $(l_{max} + \frac{1}{2})^{-3}$	Error[a]
$H_2 + F_2 \rightarrow 2\ HF$	-135.06	-134.41	0.65
$N_2 + 3\ H_2 \rightarrow 2\ NH_3$	-38.12	-38.91	-0.79
$H_2 + CO \rightarrow H_2CO$	-5.50	-5.15	0.35
$C_2H_2 + 3\ H_2 \rightarrow 2\ CH_4$	-106.53	-106.80	-0.27
$H_2CO + 2\ H_2 \rightarrow CH_4 + H_2O$	-59.81	-59.55	0.25
$HCN + 3\ H_2 \rightarrow CH_4 + NH_3$	-76.29	-76.49	-0.20
$CO_2 + 4\ H_2 \rightarrow CH_4 + 2\ H_2O$	-58.44	-58.23	0.20

[a] Error in new CBS; RMS error and MAD equal to 0.45 and 0.39 kcal/mol, respectively.

MP2 correlation energies (Table 4.6), and the higher-order contributions to the correlation energy (Table 4.7), we can now combine these components to obtain total electronic energies. There are many plausible combinations of basis sets and extrapolation procedures that must ultimately be explored. Efficient methods should use smaller basis sets for the CCSD(T) component than for the SCF and MP2 ones. The use of intermediate basis sets for the MP4(SDQ) component should also be explored, since we found this effective for the CBS-QB3 model (Table 4.2).

As a first try, we have elected to follow our treatment of the SCF and second-order correlation energies described above, and employ Eq. (6.2) to provide a linear extrapolation of the cc-pVDZ and cc-pVTZ total CBS-CCSD(T) energies obtained with Eq. (2.2), including the interference correction. These total energies reproduce the CCSD(T) limits estimated by Martin [55] via an $(l_{max} + \frac{1}{2})^{-3}$ extrapolation of the CCSD(T)/cc-pVDZ, TZ, QZ, 5Z, and 6Z basis sets to within 0.96 kcal/mol RMS error. The agreement with Martin's energies for a small set of chemical reactions is even better (Table 4.8). The use of the cc-pVnZ basis sets for PNO-$(l_{max} + \frac{1}{2})^{-3}$ double extrapolations is indeed promising.

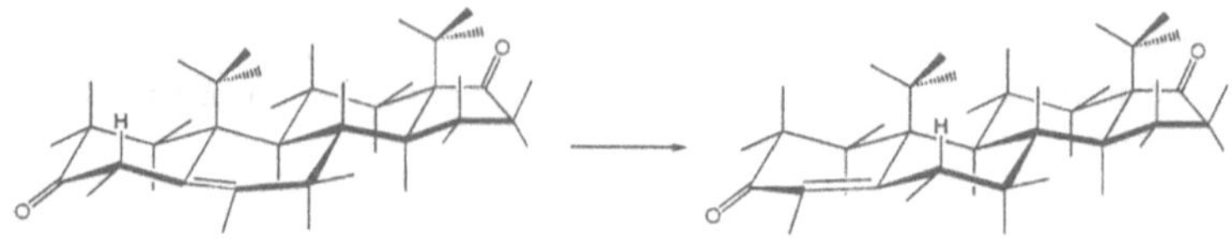

Figure 4.10 Δ^5-ketosteroid isomerase-catalyzed conversion of Δ^5-androstene-3,17-dione to the Δ^4 isomer.

8. ENZYME KINETICS AND MECHANISM

Application of CBS extrapolations to the Δ^5-ketosteroid isomerase-catalyzed conversion of Δ^5-androstene-3,17-dione to the Δ^4 isomer (Fig. 4.10) provides a test case for extensions to enzyme kinetics. This task requires integration of CBS extrapolations into multilayer ONIOM calculations [56, 57] of the steroid and the active site combined with a polarizable continuum model (PCM) treatment of bulk dielectric effects [58-60]. The goal is to reliably predict absolute rates of enzyme-catalyzed reactions within an order of magnitude, in order to verify or disprove a proposed mechanism.

Deuterium substitution for the migrating 4β proton demonstrates that the enzyme transfers it by a stereospecific intramolecular path

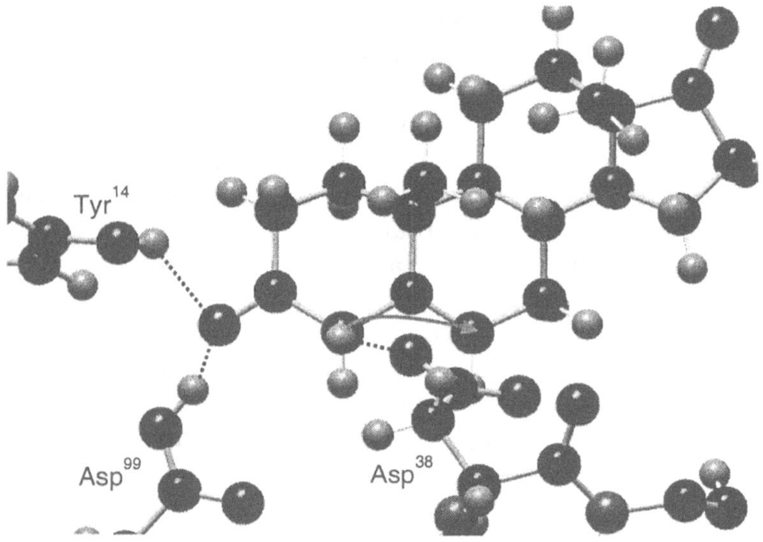

Figure 4.11 The active site of Δ^5-ketosteroid isomerase binding to Δ^5-androstene-3,17-dione.

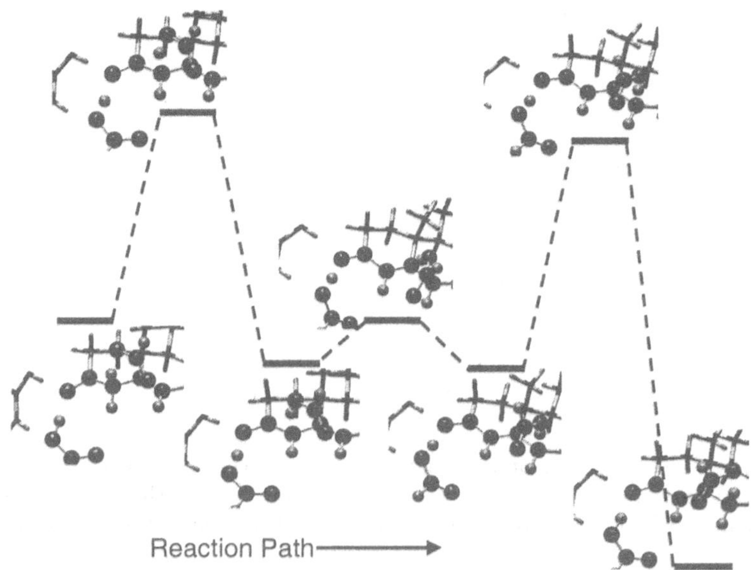

Figure 4.12 The mechanism for the isomerization of $\Delta^5 \rightarrow \Delta^4$-androstene-3,17-dione with optimized geometries for the seven key-structures.

to the 6β position without solvent exchange [61-63]. The structure of the active site (Fig. 4.11) provides a ready explanation for the observed stereochemistry. The Asp^{38} (aspartate residue 38) is well situated to escort the migrating proton, while Asp^{99} (aspartic residue 99) and the Tyr^{14} (tyrosine residue 14) stabilize the dienolate intermediate through hydrogen bonding [64].

The accepted mechanism is rather complicated [64]. First the 4β proton of the substrate transfers to Asp^{38}, then the (now protonated) carboxyl group of Asp^{38} rotates about the CH_2–$CO_2\mathbf{H}$ bond to position the migrating proton over carbon 6 of the substrate, and finally the migrating proton transfers to the 6β position on the Δ^4-androstene-3,17-dione product. The seven stationary points on the potential energy surface are therefore the reactant, two intermediates, the product, and three transition states (Fig. 4.12). Before we can calculate energies and reaction rates, we must first locate these structures on the potential energy surface, and then determine by a frequency calculation whether each stationary point is a local minimum, a first-order saddle point, or a structure that is not relevant to the reaction mechanism. These preliminary structure calculations verify whether the proposed mecha-

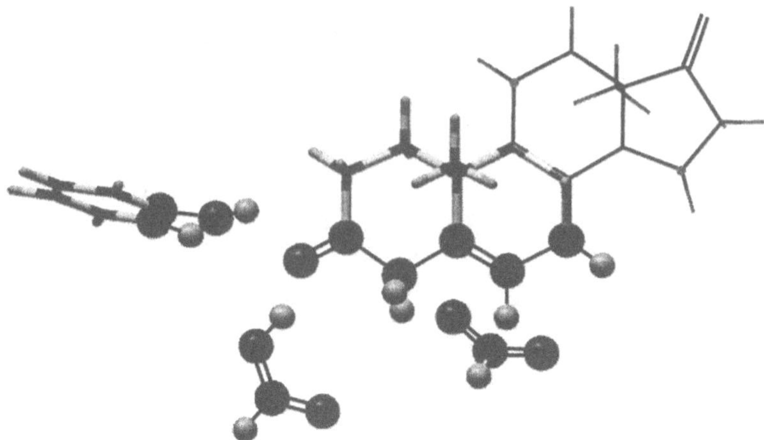

Figure 4.13 The partition of the enzyme-substrate complex into high- (ball and bond), medium- (tube), and low-level (stick) regions. Note that the aspartic acid and tyrosine residues are represented by formic acid and phenol, respectively.

nism is qualitatively consistent with the potential energy surface for our quantum-mechanical model problem.

The key functional groups for the active site are widely separated in the amino acid sequence (14, 38, and 99), as is frequently the case. We therefore elected to omit the intervening residues from our calculations entirely, and instead rely on optimization of the transition state structures to position these functional groups.

The three levels of theory selected for the preliminary ONIOM calculation [57] were: CBS-4M, HF/3-21G(*), and MM/UFF [65, 66]. Note that HF/3-21G(*) is the level of theory employed in the geometry and frequency calculations for the CBS-4M//HF/3-21G(*) compound model. Thus, the first task was the optimization of the structures for the seven stationary points on the potential energy surface using a two-level ONIOM calculation employing HF/3-21G(*) for both the high-level and the medium-level regions from the planned CBS ONIOM single-point calculations. The partition of the enzyme-substrate complex into high-, medium-, and low-level regions is indicated in Fig. 4.13. The first transition state structure was easily found with the QST2 procedure of Schlegel and co-workers [67]. However, the others proved more problematic.

The optimum structure for the first transition state placed the Asp^{99} residue (i.e. our formic acid) reasonably close to the position in which it is found in the enzyme-inhibitor crystal structure [64]. However, this functional group is not at all rigid in our model problem. It

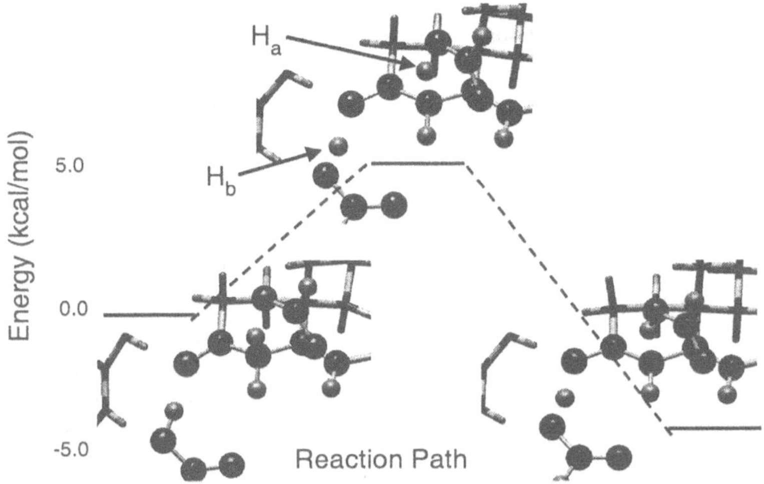

Figure 4.14 The reactant, the first transition state, and the initial enolate. The 4-β hydrogen (H_a) is transferred to Asp[38], while the carboxylic acid hydrogen (H_b) from Asp[99] is partially transferred to the 3-carbonyl oxygen.

shifts toward the 3-carbonyl in the fully optimized reactant and shifts in the opposite direction as we proceed to the product. We therefore constrained the CH hydrogen of this formic acid to be equidistant from carbons 4 and 6 of the androstene at the optimum distance for the first transition state. This emulated a rigid enzyme active site equally adept at catalyzing both proton transfer reactions. No other constraints were used in the optimization of the seven stationary points of the potential energy surface (Fig. 4.12). The Tyr[14] and Asp[99] residues, and the substrate were completely free to move and adjust their geometries as the reaction proceeded.

The optimized structures for the reactant, the first transition state, and the first enolate intermediate (Fig. 4.14) illustrate the progress along the reaction path and the nature of the intermediate (Table 4.9). The bond between the 4β hydrogen and carbon 4 lengthens from 1.122 Å to 1.903 Å as the bond order decreases from 0.853 to 0.065. Meanwhile, the distance from this hydrogen to the oxygen of Asp[38] decreases from 1.794 Å to 0.991 Å and the bond order increases from 0.370 to 0.935 as the O–H bond of the intermediate is formed. The O–H bond length of the carboxylic acid from Asp[99] increases from 0.999 Å to 1.082 Å as the bond order decreases from 0.847 to 0.636. At the same time the distance from this hydrogen to the C-3 carbonyl oxygen decreases from 1.617 Å to 1.364 Å as this bond order increases from 0.292 to 0.489. The intermediate would seem to be best described as a partially protonated

Table 4.9 Changes in bond lengths (Å) and orders.

Parameter	Bond	Reactant	TS1	Enol1
Bond length	C–H_a	1.122	1.329	1.903
	H_a–O	1.794	1.361	0.991
	OCO–H_b	0.999	1.014	1.082
	H_b–O=	1.617	1.548	1.364
Bond order	C–H_a	0.853	0.430	0.065
	H_a–O	0.370	0.631	0.935
	OCO–H_b	0.847	0.793	0.636
	H_b–O=	0.292	0.350	0.489

Table 4.10 Convergence of thermochemistry with level of theory and basis set.

Energy contribution (basis set)	Reactant	TS1	Enol1
E_{HF}(3-21G)	0.0	3.27	-10.29
E_{HF}(6-31G)	0.0	9.05	-1.77
E_{HF}[6-31+G(d',p')]	0.0	14.05	3.67
E_{HF}[6-311+G(2df,p)]	0.0	13.71	1.93
$E^{(2)}$[6-31+G(d',p')]	0.0	-10.23	-8.28
$E^{(2)}$(CBS)	0.0	-11.20	-7.33
$E^{(3)}$(6-31G)+$E^{(4)}$(6-31G)	0.0	2.17	1.54
$E^{(3)}$(CBS)+$E^{(4)}$(CBS)	0.0	2.44	1.06
E(CBS-4M)[a]	0.0	5.21	-4.18

[a] E(CBS-4M) = E_{HF}[6-311+G(2df,p)] + $E^{(2)}$(CBS) + $E^{(3)}$(CBS) + $E^{(4)}$(CBS) + E_{mp}; see Table 4.2.

enolate ion. The negative charge that was localized on the carboxylate of Asp[38] in the reactant has been delocalized over the dienolate and Asp[99] in the intermediate.

We performed CBS-4M single point energy calculations at these stationary points. The barrier height for the first proton transfer and the relative energy of the first dienolate are quite sensitive to the level of theory and basis set employed (Table 4.10 and Fig. 4.14). The initial Asp[99] (i.e. formate ion) carries the full negative charge in our model and is thus more sensitive to the Hartree-Fock basis set. Hence, the calculated Hartree-Fock barrier increases as the basis set is improved. The correlation energy favors the transition state, as is generally the case. The effects on the enolate are intermediate. The modest changes from the CBS extrapolations suggest convergence to within 1 or 2 kcal/mol. Note that the total barrier height is the sum of the Hartree-Fock and correlation energy contributions, e.g. the MP2/6-31+G(d',p') barrier height is the sum of 14.05 kcal/mol from the Hartree-Fock component and -10.23 kcal/mol from the second-order component, giving a total of 3.82 kcal/mol for the MP2 barrier.

Refinements of these calculations would include a CBS-QB3 study of the dienolate and Asp^{38} portion of our model. Since we have only used a two-layer ONIOM calculation for the geometry optimizations, a refined geometry optimization with a B3LYP/6-311G(d,p) treatment of this region would be straightforward. We should also use the IRCMax procedure [40] to determine the CBS transition state geometry and energy. Such refinements should be a part of any quantitative CBS study of enzyme kinetics. Nevertheless, the preliminary results listed in Table 4.10 clearly demonstrate the importance of achieving basis set convergence for both the SCF and the correlation energies. Even when the electronic structure is as simple as in these proton transfer reactions that maintain a closed-shell structure throughout the reaction path, low levels of theory can be quite misleading.

Calculations of bulk dielectric effects using the polarizable continuum model of Tomasi and co-workers [58-60] gave an increase in the barrier height to 12.5 kcal/mol with the recommended [68] dielectric constant ϵ equal to 18. This is the direction of change one would expect. The dielectric medium stabilizes the charge of the Asp^{38} residue (which is completely exposed on the surface of our model) more than the charge of the enolate (which is enclosed in the interior of our model). The calculated barrier height is now in a good agreement with the experimental rate constant, $k_{cat} = 30000\,sec^{-1}$ [64, 69], which implies a Gibbs free energy of activation $RT\ln(k_BT/k_{cat}h) = 11.3$ kcal/mol. However, such a large effect raises questions about the accuracy of these PCM corrections. Within the context of high-accuracy methods such as the CBS models, it would be prudent to interpret bulk dielectric effects larger than 1 or 2 kcal/mol as an indication that our model does not include sufficient detail in the region of the Asp^{38} residue. We conclude that a definitive computational study of the Δ^5-ketosteroid isomerase should include the entire Asp^{38} residue and probably some explicit water molecules in this region as well.

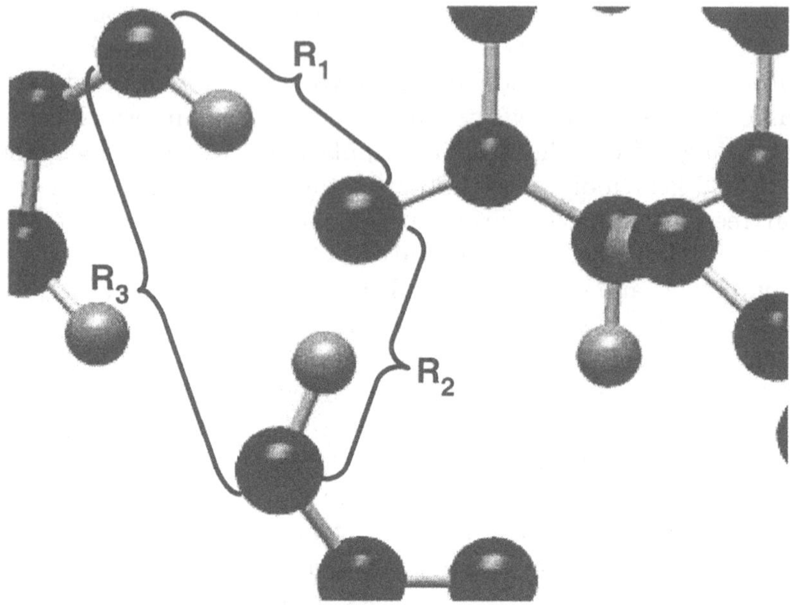

Figure 4.15 The local geometry of the Δ^5-androstene-3,17-dione in the complex with Δ^5-ketosteroid isomerase.

To close on a more positive note, we observe that the computed geometry of the enzyme-dienolate complex in the vicinity of the 3-carbonyl is insensitive to the assumed dielectric constant and is in close agreement with X-ray structures of enzyme-inhibitor complexes (see Table 4.11 and Fig. 4.15). It is really quite remarkable that 4 billion years of random walk by mother nature and a few hours of optimization with a quantum chemistry program such as Gaussian^{TM} (starting with the correct functional groups) lead to the same structure for the active

Table 4.11 Comparison of theory with experiment at the 3-carbonyl of Δ^5-androstene-3,17-dione in the complex with Δ^5-ketosteroid isomerase; see Fig. 4.15.

	R_1	R_2	R_3
$\epsilon = 0$	2.65	2.55	4.20
$\epsilon = 18$	2.57	2.54	4.19
Exp.[a]	2.62	2.56	4.06

[a] Average of 4 X-ray structures [54].

site of this enzyme. It is hard to imagine a more striking vindication of our assumption that enzymes function by reducing the energies of transition states. The mysterious ways of nature are truly beautiful [70].

9. SUMMARY

Pair natural orbital extrapolations to the complete basis set limit provide the foundation for a sequence of cost-effective CBS models. The current models: CBS-QCI/APNO, CBS-QB3, and CBS-4M, are applicable to species with 5, 10, and 20 non-hydrogen atoms, and are reliable to ca. 0.5, 1.0, and 2.0 kcal/mol respectively. These methods are applicable to transition states for chemical reactions with the IRCMax procedure. The ZC-VTST CBS-QCI/APNO model is capable of quantitative predictions of absolute rate constants. A double extrapolation promises a new generation of significantly more accurate and reliable models that will no longer require empirical corrections. The Δ^5-ketosteroid isomerase-catalyzed conversion of Δ^5- to Δ^4-androstene-3,17-dione provided a case study for the extension of such high-accuracy methods to enzyme kinetics. The impact that high-accuracy computational quantum chemistry is currently having on combustion chemistry will soon be extended to biology. This is an exciting time for computational science.

ACKNOWLEDGEMENTS

The author is grateful to the U. S. Department of Energy and Gaussian, Inc. for the continuing support of this research and recognizes the very important contributions of his many collaborators including: Marc R. Nyden, Mohammad A. Al-Laham, Joseph W. Ochterski, John A. Montgomery, Jr., David K. Malick, Michael J. Frisch, and Rex F. Pratt.

REFERENCES

1. G.A. Petersson, *Theor. Chem. Acc.* **103**, 190 (2000).
2. K. K. Irikura and D. J. Frurip (Ed.), *Computational Thermochemistry*, ACS Symposium Series 677, American Chemical Society, Washington, D. C. (1998).
3. R. J. Bartlett and G. D. Purvis, *Int. J. Quant. Chem.* **14**, 516 (1978); G. D. Purvis and R. J. Bartlett, *J. Chem. Phys.* **76**, 1910 (1982).

4. J. A. Pople, M. Head-Gordon, and K. Raghavachari, *J. Chem. Phys.* **87**, 5968 (1987); K. Raghavachari, G. W. Trucks, J. A. Pople, and M. Head-Gordon, *Chem. Phys. Lett.* **157**, 479 (1989).
5. J. M. L. Martin and G. de Olivera, *J. Chem. Phys.* **111**, 1843 (1999).
6. C. Schwartz, *Phys. Rev.* **126**, 1015 (1962).
7. C. Schwartz, in *Methods in Computational Physics*, Vol.2, B. Alder, S. Fernbach, M. Rotenberg (Eds.), Academic, New York (1963).
8. M. R. Nyden and G. A. Petersson, *J. Chem. Phys.* **75**, 1843 (1981).
9. G. A. Petersson and M. R. Nyden, *J. Chem. Phys.* **75**, 3423 (1981).
10. G.A. Petersson and S. L. Licht, *J. Chem. Phys.* **75**, 4556 (1981).
11. F. B. Brown and D. G. Truhlar, *Chem. Phys. Lett.* **117**, 307 (1985).
12. T. H. Dunning, Jr., *J. Chem. Phys.* **90**, 1007 (1989).
13. P. E. M. Siegbahn, M. R. A. Blomberg, and M. Svensson, *Chem. Phys. Lett.* **223**, 35 (1994).
14. P. O. Löwdin, *Phys. Rev.* **97**, 1474 (1955).
15. F. W. Byron, Jr., and C. J. Joachain, *Phys Rev.* **146**, 1 (1966).
16. F. W. Byron, Jr., and C. J. Joachain, *Phys Rev.* **157**, 7 (1967).
17. C. Møller and M. S. Plesset, *Phys Rev.* **46**, 618 (1934).
18. C. F. Bunge, *Theor. Chim. Acta (Berlin)* **16**, 126 (1970).
19. K. Jankowski and P. Malinowski, *Chem. Phys. Lett.* **54**, 68 (1978); K. Jankowski, P. Malinowski, and M. Polasik, *J. Phys.* **B 12**, 3157 (1979); K. Jankowski, and P. Malinowski, *Phys. Rev.* **A 21**, 45 (1980); K. Jankowski, P. Malinowski, and M. Polasik, *Phys. Rev.* **A 21**, 51 (1980).
20. J. W. Ochterski, G. A. Petersson, and J. A. Montgomery, Jr., *J. Chem. Phys.* **104**, 2598 (1996).
21. G. A. Petersson and M. Braunstein, *J. Chem. Phys.* **83**, 5129 (1985).
22. J. A. Montgomery, Jr., J. W. Ochterski, and G. A. Petersson, *J. Chem. Phys.* **101**, 5900 (1994).
23. J. A. Montgomery, Jr., M. J. Frisch, J. W. Ochterski, and G. A. Petersson, *J. Chem. Phys.* **110**, 2822 (1999).
24. J. A. Montgomery, Jr., M. J. Frisch, J. W. Ochterski, and G. A. Petersson, *J. Chem. Phys.* **112**, 6532 (2000).
25. Gaussian 98, M. J. Frisch, G. W. Trucks, H. B. Schlegel, G. E. Scuseria, M. A. Robb, J. R. Cheeseman, V. G. Zakrzewski, J. A. Montgomery, Jr., R. E. Stratmann, J. C. Burant, S. Dapprich, J. M. Millam, A. D. Daniels, K. N. Kudin, M. C. Strain, O. Farkas, J. Tomasi, V. Barone, M. Cossi, R. Cammi, B. Mennucci, C. Pomelli, C. Adamo, S. Clifford, J. Ochterski, G. A. Petersson, P. Y. Ayala, Q. Cui, K. Morokuma, D. K. Malick, A. D. Rabuck, K. Raghavachari, J. B. Foresman, J. V. Ortiz, A. G. Baboul, J. Cioslowski, B. B. Stefanov, G. Liu, A. Liashenko, P. Piskorz, I. Komaromi, R. Gomperts, R. L. Martin, D. J. Fox, T. Keith, M. A. Al-Laham, C. Y. Peng, A. Nanayakkara, M. Challacombe, P. M. W. Gill, B. Johnson, W. Chen, M. W. Wong, J. L. Andres, C. Gonzalez, M. Head-Gordon, E. S. Replogle, and J. A. Pople, Gaussian, Inc., Pittsburgh PA, 1998.

26. Improvements in hardware and software now make QCISD/6-311G** frequencies practical for the CBS-QCI/APNO model.
27. L. A. Curtiss, C. Jones, G. W. Trucks, K. Raghavachari, and J. A. Pople, J. *Chem. Phys.* **93**, 2537 (1990).
28. P. Pulay, in *Applications of Electronic Structure Theory*, H. F. Schaefer, III (Ed.), Plenum, New York (1977), p. 153.
29. J. A. Pople, R. Krishnan, H. B. Schlegel, and J. S. Binkley, *Int. J. Quant. Chem. Symp.* **13**, 325 (1979).
30. H. B. Schlegel, *J. Comput. Chem.* **3**, 214 (1982).
31. N. C. Handy and H. F. Schaefer, III, *J. Chem. Phys.* **81**, 5031 (1984).
32. H. B. Schlegel, J. S. Binkley, and J. A. Pople, *J. Chem. Phys.* **80**, 1976 (1984).
33. M. J. Frisch, M. Head-Gordon, and J. A. Pople, *Chem. Phys. Lett.* **166**, 275; (1990); M. J. Frisch, M. Head-Gordon, and J. A. Pople, *Chem. Phys. Lett.* **166**, 281 (1990)
34. B. G. Johnson and M. J. Frisch, *J. Chem. Phys.* **100**, 7429 (1994).
35. M. Head-Gordon and T. Head-Gordon, *Chem. Phys. Lett.* **220**, 122 (1994).
36. C. Peng and H. B. Schlegel, *Israel J. Chem.* **33**, 449 (1994).
37. D. G. Truhlar and A. Kuppermann, *J. Am. Chem. Soc.* **93**, 1840 (1971).
38. C. Gonzalez and H. B. Schlegel, *J. Phys. Chem.* **90**, 2154 (1989).
39. J. C. Corchado, E J. Olivares del Valle, and J. Espinosa-Garcia, *J. Phys. Chem.* **97**, 9129 (1993).
40. D. K. Malick, G. A. Petersson, and J. A. Montgomery, Jr., *J. Chem. Phys.* **108**, 5704 (1998).
41. J. B. Foresman and Æ. Frisch, *Exploring Chemistry with Electronic Structure Methods, 2nd ed.*, Gaussian, Inc., Pittsburgh, PA (1996).
42. D. G. Truhlar, *J. Chem. Phys.* **53**, 2041 (1970).
43. B. C. Garrett and D. G. Truhlar, *J. Phys. Chem.* **83**, 1052 (1979); B. C. Garrett and D. G. Truhlar, *J. Phys. Chem.* **83**, 1079 (1979).
44. *NIST Chemical Kinetics Database, Version 6.0* (1998).
45. W. Kutzelnigg, *Theor. Chim. Acta* **68**, 445 (1985); W. Klopper and W. Kutzelnigg, *Chem. Phys. Lett.* **134**, 17 (1987); W. Klopper and W. Kutzelnigg, *Stud. Phys. Theor. Chem.* **62**, 45 (1989).
46. J. Noga, W. Klopper, and W. Kutzelnigg, in *Recent Advances in Coupled Cluster Methods* R. J. Bartlett (Ed.), World Scientific, Singapore (1997), p. 1.
47. W. Klopper, in *The Encyclopedia of Computational Chemistry*, P. v. R. Schleyer, et al. (Eds.), Wiley, Chichester, (1998), p. 2351.
48. W. Klopper, K. L. Bak, P. Jørgensen, J. Olsen, and T. Helgaker, *J. Phys.* **B 32**, R103 (1999).
49. R. A. Kendall, T. H. Dunning, Jr., and R. J. Harrison, *J. Chem. Phys.* **96**, 6796 (1992); D. E. Woon and T. H. Dunning, Jr., *J. Chem. Phys.* **98**, 1358 (1993); D. E. Woon and T. H. Dunning, Jr., *J. Chem. Phys.* **100**, 2975 (1994); D. E. Woon and T. H. Dunning, Jr., *J. Chem. Phys.* **103**, 4572 (1995); A. K. Wilson, T. van Mourik, and T. H. Dunning, Jr., *J. Mol. Struct. (Theochem)* **338**, 339 (1996);

D. E. Woon, K. A. Peterson, and T. H. Dunning, Jr., *J. Chem. Phys.* **109**, 2233 (1998).

50. G. A. Petersson and M. J. Frisch, *J. Phys. Chem.* **104**, 2183 (2000).
51. D. Feller, *J. Chem. Phys.* **96**, 6104; D. Feller, *J. Chem. Phys.* **98**, 7059 (1993).
52. K. A. Peterson, D. E. Woon, and T. H. Dunning, Jr., *J. Chem. Phys.* **100**, 7410 (1994).
53. A. K. Wilson and T.H. Dunning, Jr., *J. Chem. Phys.* **106**, 8718 (1997).
54. F. B. van Duijneveldt, *IBM Publ. RI 945*, Yorktown Hts., New York (1971).
55. J. M. L. Martin, *Theor. Chem. Acc.* **97**, 227 (1997).
56. M. Svensson, S. Humbel, R. D. J. Froese, T. Matsubara, S. Sieber, and K. Morokuma, *J. Phys. Chem.* **100**, 19357 (1996).
57. T. Vreven and K. Morokuma, *J. Chem. Phys.* **113**, 2969 (2000).
58. M. Cossi, V. Barone, R. Cammi, and J. Tomasi, *Chem. Phys. Lett.* **255**, 327 (1996).
59. V. Barone, M. Cossi, and J. Tomasi, *J. Chem. Phys.* **107**, 3210 (1997).
60. V. Barone, M. Cossi, and J. Tomasi, *J. Comp. Chem.* **19**, 404 (1998).
61. T. C. Eames, D. C. Hawkinson, and R. M. Pollack, *J. Am. Chem. Soc.* **112**, 1996 (1990).
62. L. Xue, A Kuliopulos, A. S. Mildvan, and P. Talalay, *Biochemistry* **30**, 4991 (1991).
63. C. M. Holman and W. F. Bebisek, *Biochemistry* **33**, 2672 (1994).
64. G. Choi, N. C. Ha, S. W. Kim, D. H. Kim, S. Park, B. H. Oh, and K. Y. Choi, *Biochemistry* **39**, 903 (2000).
65. A. K. Rappé, C. J. Casewit, K. S. Colwell, W. A. Goddard III and W. M. Skiff, *J. Am. Chem. Soc.* **114**, 10024 (1992).
66. A. K. Rappé and W. A. Goddard III, *J. Phys. Chem.* **95**, 3358 (1991).
67. C. Peng and H. B. Schlegel, *Israel J. Chem.* **33**, 449 (1994).
68. K. Oh, S. Cha, D. Kim, H. Cho, N. Ha, G. Choi, J. Lee, P. Tarakeshwar, H. Son, K. Choi, B. Oh, and K. Kim, *Biochemistry* **39**, 13891 (2000).
69. J. M. Schwab and B. S. Henderson, *Chem. Rev.* **90**, 1203 (1990).
70. "The most beautiful thing we can experience is the mysterious. It is the source of all true art and science." Albert Einstein in *The Expanded Quotable Einstein*, collected and edited by A. Calaprice, Princeton Univ. Press, Princeton, N. J. (2000).

Chapter 5

Application and Testing of Diagonal, Partial Third-Order Electron Propagator Approximations

Antonio M. Ferreira, Gustavo Seabra, O. Dolgounitcheva,
V. G. Zakrzewski, and J. V. Ortiz
Department of Chemistry, Kansas State University, Manhattan, Kansas 66506-3701

1. INTRODUCTION

Ionization energies and electron affinities are among the most often sought thermochemical data. The importance of electron binding energies is reflected by their presence in a variety of thermodynamic arguments, including thermochemical cycles of acidity and basicity, complexation energies, and oxidation-reduction reactions. Many spectroscopic methods founded on the photoelectric effect, mass spectrometry, electron scattering, and other techniques measure ionization energies and electron affinities. The precision of these experiments in measuring transition energies often contrasts with the paucity of information they generate on accompanying molecular and ionic structures. Computational means of estimating ionization energies and electron affinities therefore provide indispensable corroborative information on structures, especially as the scope of thermochemical and spectroscopic measurements expands.

Given the ubiquitous character of molecular orbital concepts in contemporary discourse on electronic structure, ionization energies and electron affinities provide valuable parameters for one-electron models of chemical bonding and spectra. Electron binding energies may be assigned to delocalized molecular orbitals and thereby provide measures of chemical reactivity. Notions of hardness and softness, electronegativity,

J. Cioslowski (ed.), Quantum-Mechanical Prediction of Thermochemical Data, 131–160.

and other qualitative concepts often appeal to molecular orbitals and their corresponding energies.

While many experimental techniques and the majority of computational strategies focus on the generation of increasingly precise ionization energies and electron affinities, fewer methods emphasize the connection between these electron binding energies and the changes in electronic structure they represent. Because one-electron concepts have a history of generating powerful ordering principles for the formulation of hypotheses about electronic structure, it is desirable to use theoretical techniques that show how to connect electron binding energies to orbitals.

2. ELECTRON PROPAGATOR CONCEPTS

Electron propagator theory [1-11] provides a conceptual and computational foundation for this path of inquiry. First, this theory, which is also known as one-electron Green's function theory or as the equation-of-motion method, provides a rigorous framework for calculations of ionization energies and electron affinities. Second, to each electron binding energy ε_{p}, electron propagator theory associates a function of the coordinates of a single electron $\phi_{\mathrm{p}}(\mathbf{x})$. Both of these objects are results of solving a pseudoeigenvalue problem,

$$\hat{\mathrm{H}}^{\mathrm{eff}}\phi_{\mathrm{p}}(\mathbf{x}) = \varepsilon_{\mathrm{p}}\,\phi_{\mathrm{p}}(\mathbf{x})\,. \tag{2.1}$$

A special case of this approach is represented by the Hartree-Fock equations, where the effective operator $\hat{\mathrm{H}}^{\mathrm{eff}}$ contains the usual kinetic ($\hat{\mathrm{T}}$), nuclear attraction ($\hat{\mathrm{U}}$), Coulomb ($\hat{\mathrm{J}}$), and exchange ($\hat{\mathrm{K}}$) components such that

$$\hat{\mathrm{H}}^{\mathrm{eff}} = \hat{\mathrm{F}} = \hat{\mathrm{T}} + \hat{\mathrm{U}} + \hat{\mathrm{J}} - \hat{\mathrm{K}}\,. \tag{2.2}$$

Since the $\hat{\mathrm{J}}$ and $\hat{\mathrm{K}}$ operators depend on the occupied orbitals, the pseudoeigenvalue problem must be solved iteratively until consistency is achieved between orbitals that determine $\hat{\mathrm{J}}$ and $\hat{\mathrm{K}}$ and those that emerge as eigenfunctions of $\hat{\mathrm{H}}^{\mathrm{eff}}$, which in this approximation is known as the Fock operator $\hat{\mathrm{F}}$.

Electron propagator formalism allows for generalizations that include the effects of correlation. Here the pseudoeigenvalue problem has the following structure

$$\left[\hat{\mathrm{F}} + \hat{\Sigma}(\varepsilon_{\mathrm{p}})\right]\phi_{\mathrm{p}}(\mathbf{x}) = \varepsilon_{\mathrm{p}}\,\phi_{\mathrm{p}}(\mathbf{x})\,. \tag{2.3}$$

Now the Fock operator is supplemented by the self-energy operator $\hat{\Sigma}(\mathrm{E})$. This operator depends on an energy parameter E and is nonlocal. All orbital relaxation effects between initial and final states may be included in the self-energy operator, as well as all differences in the correlation energies of these states. As in the Hartree-Fock case, matrix elements of the Fock operator still depend on the charge-bond order density matrix **D** (also known as the one-electron density matrix) according to

$$F_{rs} = T_{rs} + U_{rs} + \sum_{tu} \langle rt||su\rangle D_{tu} , \tag{2.4}$$

but **D** may pertain to a correlated reference state. The energy dependence of the correlated effective operator $\hat{H}^{eff}$, where

$$\hat{H}^{eff}(E) = \hat{F} + \hat{\Sigma}(E) , \tag{2.5}$$

indicates that the correlated pseudoeigenvalue problem must also contain iterations with respect to E. A search for electron binding energies requires that a guess energy be inserted into $\hat{H}^{eff}(E)$, leading to new eigenvalues which may be reinserted into $\hat{H}^{eff}(E)$ in a cyclic manner until consistency is obtained between the operator and its eigenvalues. Approximations to $\hat{\Sigma}(E)$ may be systematically extended until, in principle, exact ionization energies and electron affinities emerge as $\{\varepsilon_p\}$ values.

Eigenfunctions that accompany these eigenvalues have a clear physical meaning that corresponds to electron attachment or detachment. These functions are known as Dyson orbitals, Feynman-Dyson amplitudes, or generalized overlap amplitudes. For ionization energies, they are given by

$$\begin{aligned} \phi_p(\mathbf{x}_1) &= N^{1/2} \int \Psi_N(\mathbf{x}_1, \mathbf{x}_2, \mathbf{x}_3, \ldots, \mathbf{x}_N)\, \Psi^*_{N-1,p}(\mathbf{x}_2, \mathbf{x}_3, \mathbf{x}_4, \ldots, \mathbf{x}_N) \\ &\quad \times\ d\mathbf{x}_2 d\mathbf{x}_3 d\mathbf{x}_4 \ldots d\mathbf{x}_N , \end{aligned} \tag{2.6}$$

where $\mathbf{x}_i$ is the space-spin coordinate of electron i. The Dyson orbital corresponding to the energy difference between the N-electron state Ψ_N and the p-th electron-detached state $\Psi_{N-1,p}$ may be used to calculate cross sections for various types of photoionization and electron scattering processes. For example, photoionization intensities $\{I_p\}$ may be determined via

$$I_p = \kappa\, |\langle \phi_p | \nabla \chi \rangle|^2 , \tag{2.7}$$

where χ is a description of the ejected photoelectron. For electron affinities, the formula for the Dyson orbital reads

$$\begin{aligned} \phi_p(\mathbf{x}_1) &= (N+1)^{1/2} \int \Psi_{N+1,p}(\mathbf{x}_1, \mathbf{x}_2, \mathbf{x}_3, \ldots, \mathbf{x}_N, \mathbf{x}_{N+1}) \\ &\quad \times\ \Psi^*_N(\mathbf{x}_2, \mathbf{x}_3, \mathbf{x}_4, \ldots, \mathbf{x}_N, \mathbf{x}_{N+1}) \\ &\quad \times\ d\mathbf{x}_2 d\mathbf{x}_3 d\mathbf{x}_4 \ldots d\mathbf{x}_N d\mathbf{x}_{N+1} . \end{aligned} \tag{2.8}$$

In the Hartree-Fock, frozen-orbital case, the reference state consists of a single determinant of spinorbitals and the final states differ by the addition or subtraction of an electron in a canonical spinorbital. The overlaps between states of unequal numbers of electrons represented by the Dyson orbital formulae reduce to occupied or virtual orbitals which are solutions of the canonical Hartree-Fock equations. Dyson orbitals may also be obtained from configuration interaction wavefunctions. Electron propagator calculations, however, avoid the evaluation of complicated many-electron wavefunctions (and their energies) in favor of direct evaluation of electron binding energies and their associated Dyson orbitals. Note that for correlated calculations, the Dyson orbitals are not necessarily normalized. The pole strength P is given by

$$P_p = \int |\phi_p(x)|^2 \, dx . \tag{2.9}$$

In the Hartree-Fock, frozen-orbital case, P_p acquires its maximum value, unity. Final states with large correlation effects are characterized by low pole strengths. Transition intensities, such as those in Eq. (2.7), are proportional to P_p.

3. AN ECONOMICAL APPROXIMATION: P3

Canonical Hartree-Fock orbital energies are a convenient and powerful foundation for estimating the smallest vertical electron binding energies of closed-shell molecules. This approximation, which is based on Koopmans's theorem, is the most often used method for assigning the lowest peaks in photoelectron spectra. However, there are many classes of important molecules for which the Koopmans approximation fails to predict the correct order of final states. Average errors made by this frozen-orbital, uncorrelated method are between 1 and 2 eV for valence ionization energies. More confident assignments require that these errors be reduced.

Perturbative expressions for the self-energy operator can achieve this goal for large, closed-shell molecules. In this review, we will concentrate on an approximation developed for this purpose, the partial third-order, or P3, approximation. P3 calculations have been carried out for a variety of molecules. A tabulation of these calculations is given in Table 5.1.

The original derivation of the P3 method was accompanied by test calculations on challenging, but small, closed-shell molecules with various basis sets [12]. The average absolute error was approximately 0.2 eV

Table 5.1 Recent applications of the P3 method.

Reference molecule or ion	Year	Ref.
borazine	1996	12
azabenzenes	1996	13
dichlorobenzene	1996	14
anthracene, phenanthrene, and naphthacene	1996	15
chlorobenzene	1996	16
sym-tetrazine	1997	17
carbon quadranions	1997	18
small anions	1997	19
acridine, phenazine, and diazaphenanthrene	1997	20
benzopyrenes	1997	21
C_7^{2-}	1998	22
anisole and thioanisole	1998	23
butadiene	1999	24
uracil and adenine	2000	25
naphthalene	2000	26
guanine	2000	27
dicarboxylate dianions	2000	28
double-Rydberg anions	2000	29

for vertical ionization energies below 20 eV. Since 1996, the P3 method has been applied chiefly to the ionization energies of organic molecules. For nitrogen-containing heterocycles, P3 corrections to Koopmans results are essential in making assignments of photoelectron spectra. Correlation corrections generally are much larger for hole states with large contributions from nonbonding, nitrogen-centered functions than for delocalized π levels. Therefore, P3 results often produce a different ordering of the cationic states. The accuracy of P3 predictions generally suffices to make reliable assignments. Several reviews on electron propagator theory have discussed relationships between P3 and other methods [9-11].

The P3 method is generally implemented in the diagonal self-energy approximation. Here, off-diagonal elements of the self-energy matrix in the canonical, Hartree-Fock orbital basis are set to zero. In the P3 approximation, correlation contributions to the Fock matrix (also known as the energy-independent, or constant, part of the self-energy matrix) are ignored. The pseudoeigenvalue problem therefore reduces to separate equations for each canonical, Hartree-Fock orbital:

$$F_{pp} + \Sigma_{pp}(E) = E \tag{3.1}$$

or

$$\varepsilon_p^{HF} + \Sigma_{pp}(E) = E. \tag{3.2}$$

Only energy iterations are needed in the diagonal self-energy approximation. For example, $\Sigma_{pp}(E)$ may be evaluated at $E = \varepsilon_p^{HF}$ to obtain a new guess for E. The latter value is reinserted into $\Sigma_{pp}(E)$ and the process continues until consecutive energy guesses agree to within 0.01 μE_h of each other. Alternatively, one may use Newton's method for solving the roots of a complicated function such that

$$E - \varepsilon_p^{HF} - \Sigma_{pp}(E) = 0. \tag{3.3}$$

This procedure requires analytical expressions for $\Sigma_{pp}(E)$ and its derivative with respect to E; it usually converges in three iterations. Neglect of off-diagonal elements of the self-energy matrix also implies that the corresponding Dyson orbital is given by

$$\phi_p(\mathbf{x}) = P_p^{1/2}\, \phi_p^{HF}(\mathbf{x}) \ , \tag{3.4}$$

where the pole strength P_p is determined by

$$P_p = \left[1 - \frac{d\Sigma_{pp}(E)}{dE}\right]^{-1} . \tag{3.5}$$

In the latter expression, the derivative is evaluated at the converged energy. Diagonal self-energy approximations therefore subject a frozen Hartree-Fock orbital $\phi_p^{HF}(\mathbf{x})$ to an energy-dependent correlation potential $\Sigma_{pp}(E)$.

Diagonal matrix elements of the P3 self-energy approximation may be expressed in terms of canonical Hartree-Fock orbital energies and electron repulsion integrals in this basis. For ionization energies, where the index p pertains to an occupied spinorbital in the Hartree-Fock determinant,

$$\begin{aligned} \Sigma_{pp}^{P3}(E) &= \frac{1}{2}\sum_{iab} \frac{\langle pi||ab\rangle\langle ab||pi\rangle}{E+\varepsilon_i-\varepsilon_a-\varepsilon_b} + \frac{1}{2}\sum_{aij} \frac{W_{paij}\,\langle pa||ij\rangle}{E+\varepsilon_a-\varepsilon_i-\varepsilon_j} \\ &+ \frac{1}{2}\sum_{aij} \frac{U_{paij}(E)\,\langle ij||pa\rangle}{E+\varepsilon_a-\varepsilon_i-\varepsilon_j} \ , \end{aligned} \tag{3.6}$$

where

$$W_{paij} = \langle pa||ij\rangle + \frac{1}{2}\sum_{bc}\frac{\langle pa||bc\rangle\langle bc||ij\rangle}{\varepsilon_i+\varepsilon_j-\varepsilon_b-\varepsilon_c} + (1-P_{ij})\sum_{bk}\frac{\langle pk||bi\rangle\langle ba||jk\rangle}{\varepsilon_j+\varepsilon_k-\varepsilon_a-\varepsilon_b} \quad (3.7)$$

and

$$U_{paij}(E) = -\frac{1}{2}\sum_{kl}\frac{\langle pa||kl\rangle\langle kl||ij\rangle}{E+\varepsilon_a-\varepsilon_k-\varepsilon_l} - (1-P_{ij})\sum_{bk}\frac{\langle pb||jk\rangle\langle ak||bi\rangle}{E+\varepsilon_b-\varepsilon_j-\varepsilon_k}\,. \quad (3.8)$$

Indices i, j, k, ... (a, b, c, ...) refer to occupied (virtual) spinorbitals. Each of the terms in Eq. (3.6) may be interpreted in terms of simple concepts. The first term pertains to pair correlation energies in the reference state that are missing in the final state due to removal of an electron from the occupied spinorbital p. Summing these terms over all occupied p and setting $E = \varepsilon_p$ for each term would recover the second-order, perturbative correction to the Hartree-Fock total energy of the reference state. The remaining terms account for orbital relaxation and electron correlation in the final state. When either the i or j indices are equal to p, orbital relaxation is described by excitations of electrons into the now vacant, but previously (that is, in the reference determinant) occupied spinorbital p. To describe electron correlation in the final state in terms of spinorbitals optimized for the reference state, it is crucial to include the second and third terms of Eq. (3.7) as well the terms involving the U intermediates of Eq. (3.8). Note that these terms are second-order in electron interaction and therefore generate third-order terms in the self-energy matrix.

For each ionization energy of index p, evaluation of W elements requires arithmetic operations with an O^2V^3 scaling factor, where O is the number of occupied spinorbitals and V is the number of virtual spinorbitals. For each value of E, the U elements must be reevaluated, but the scaling factor here is only O^3V^2. Since V is generally much larger than O, the latter steps proceed relatively quickly. A complete set of transformed two-electron integrals is not needed, for the set where all four indices are virtual does not appear in these equations. The largest set of integrals, with one occupied and three virtual indices, is needed only in the first summation of Eq. (3.7). Efficient programs may avoid the evaluation and storage of these transformed integrals by performing this summation with semidirect algorithms [30].

For electron affinities, where the index p pertains to a virtual spinorbital,

$$\Sigma_{pp}^{P3}(E) = \frac{1}{2}\sum_{aij}\frac{\langle pa||ij\rangle\langle ij||pa\rangle}{E+\varepsilon_a-\varepsilon_i-\varepsilon_j}+\frac{1}{2}\sum_{iab}\frac{W_{piab}\,\langle pi||ab\rangle}{E+\varepsilon_i-\varepsilon_a-\varepsilon_b} + \frac{1}{2}\sum_{iab}\frac{U_{piab}(E)\,\langle ab||pi\rangle}{E+\varepsilon_i-\varepsilon_a-\varepsilon_b}\,, \quad (3.9)$$

where

$$W_{piab} = \langle pi||ab\rangle + \frac{1}{2}\sum_{jk}\frac{\langle pi||jk\rangle\langle jk||ab\rangle}{\varepsilon_j+\varepsilon_k-\varepsilon_a-\varepsilon_b} + (1-P_{ab})\sum_{jc}\frac{\langle pc||ja\rangle\langle ji||bc\rangle}{\varepsilon_i+\varepsilon_j-\varepsilon_b-\varepsilon_c} \quad (3.10)$$

and

$$U_{piab}(E) = \frac{1}{2}\sum_{cd}\frac{\langle pi||cd\rangle\langle cd||ab\rangle}{E+\varepsilon_i-\varepsilon_c-\varepsilon_d}+(1-P_{ab})\sum_{jc}\frac{\langle pj||bc\rangle\langle ic||ja\rangle}{E+\varepsilon_j-\varepsilon_b-\varepsilon_c}\,. \quad (3.11)$$

These formulae are similar to those for the ionization energy case, but with the roles of occupied and virtual indices being reversed. Interpretation of the terms proceeds in an analogous manner. The first summation in Eq. (3.11) now dominates the arithmetic and storage requirements of the calculation. Its scaling factor is OV^4 and it requires electron repulsion integrals with four virtual indices. Practical calculations generally require a semidirect algorithm for this step.

4. OTHER DIAGONAL APPROXIMATIONS

All second-order terms are retained in the P3 self-energy formulae for ionization energies and electron affinities. There are no differences between the expressions used for ionization energies and electron affinities in the second-order self-energy, which reads

$$\Sigma_{pp}^{(2)}(E) = \frac{1}{2}\sum_{aij}\frac{\langle pa||ij\rangle\langle ij||pa\rangle}{E+\varepsilon_a-\varepsilon_i-\varepsilon_j}+\frac{1}{2}\sum_{iab}\frac{\langle pi||ab\rangle\langle ab||pi\rangle}{E+\varepsilon_i-\varepsilon_a-\varepsilon_b} \quad (4.1)$$

for all p. Diagonal, second-order calculations generally overestimate the exact correlation correction to the Hartree-Fock orbital energy. The absolute values of the ensuing errors are often as large as those of Koopmans's theorem [31].

More satisfactory results are obtained from full third-order calculations [32, 33]. Diagonal elements of the full third-order, self-energy matrix are given by

$$\begin{aligned} \Sigma^{(3)}_{pp}(E) &= \sum_{aij} \frac{\left[W_{paij} + \frac{1}{2}U_{paij}(E)\right] \langle pa||ij\rangle}{E + \varepsilon_a - \varepsilon_i - \varepsilon_j} \\ &+ \sum_{iab} \frac{\left[W_{piab} + \frac{1}{2}U_{piab}(E)\right] \langle pi||ab\rangle}{E + \varepsilon_i - \varepsilon_a - \varepsilon_b} \\ &+ \Sigma^{(3)}_{pp}(\infty) \,. \end{aligned} \tag{4.2}$$

Terms containing the W intermediates no longer contain a factor of $\frac{1}{2}$. The energy-independent, third-order term, $\Sigma^{(3)}_{pp}(\infty)$, is a Coulomb-exchange matrix element determined by second-order corrections to the density matrix, where

$$\Sigma^{(3)}_{pp}(\infty) = \sum_{rs} \langle pr||ps\rangle \, D^{(2)}_{rs} \,. \tag{4.3}$$

Third-order results for closed-shell molecules have average absolute errors of 0.6 - 0.7 eV [31]. Transformed integrals with four virtual indices and OV^4 contractions for each value of E are required for the U intermediate, which is needed for ionization energy as well as electron affinity calculations.

Second-order and third-order results often bracket the true correction to $\{\varepsilon_p^{HF}\}$. Three schemes that scale the third-order terms in various ways are known as the Outer Valence Green's Function (OVGF) [8]. In OVGF calculations, one of these three recipes is chosen as the recommended one according to rules based on numerical criteria. These criteria involve quantities that are derived from ratios of various constituent terms of the self-energy matrix elements. Average absolute errors for closed-shell molecules are somewhat larger than for P3 [31].

5. NONDIAGONAL APPROXIMATIONS

For many ionization energies and electron affinities, diagonal self-energy approximations are inappropriate. Methods with nondiagonal self-energies allow Dyson orbitals to be written as linear combinations of reference-state orbitals. In most of these approximations, combinations of canonical, Hartree-Fock orbitals are used for this purpose, i.e.

$$\phi_{\mathrm{p}}^{\mathrm{Dyson}}(\mathbf{x}) = \sum_{\mathrm{q}} \mathrm{C}_{\mathrm{pq}}\, \phi_{\mathrm{q}}^{\mathrm{HF}}(\mathbf{x})\,. \tag{5.1}$$

For normalized Hartree-Fock orbitals, the pole strength reads

$$\mathrm{P}_{\mathrm{p}} = \sum_{\mathrm{q}} |\mathrm{C}_{\mathrm{pq}}|^2\,. \tag{5.2}$$

Nondiagonal self-energy approximations are usually renormalized in the sense that they contains terms in all orders of electron interaction. For example, the 2p-h Tamm-Dancoff approximation (2ph-TDA) is suitable for qualitative descriptions of correlation (shake-up) final states in inner-valence photoelectron spectra [34]. An extension of this method, known as the third-order algebraic diagrammatic construction [ADC(3)] includes all third-order terms in the self-energy matrix. While it retains the ability of 2ph-TDA to generate a simple description of correlation states, ADC(3) is competitive with OVGF in describing final states where the Koopmans picture is qualitatively valid [8]. A nondiagonal, renormalized extension of the P3 method that retains all second-order, self-energy terms is known as NR2 [35]. For valence ionization energies of closed-shell molecules, NR2 is somewhat more accurate than P3, but it is also applicable to final states with large correlation effects [26, 36]. One pays for the enhanced versatility of these methods with increased arithmetic and storage requirements. The relatively modest demands of NR2 calculations make this approximation an attractive target for algorithmic improvements.

It is also possible to employ highly correlated reference states as an alternative to methods that employ Hartree-Fock orbitals. Multiconfigurational, spin-tensor, electron propagator theory adopts multiconfigurational, self-consistent-field reference states [37]. Perturbative corrections to these reference states have been introduced recently [38].

Another approach of this kind uses the approximate Brueckner orbitals from a so-called Brueckner doubles, coupled-cluster calculation [39, 40]. Methods of this kind are distinguished by their versatility and have been applied to valence ionization energies of closed-shell molecules, electron detachment energies of highly correlated anions, core ionization

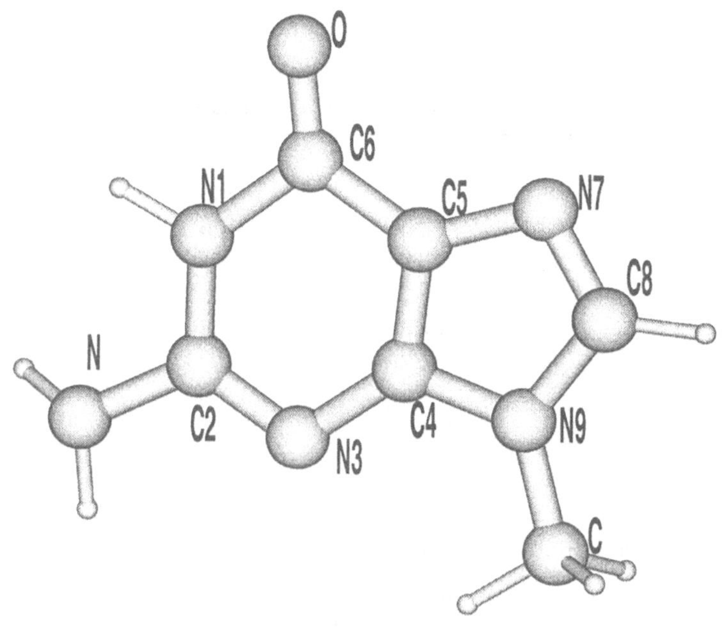

Figure 5.1 The keto form of 9-methylguanine.

energies, and photoelectron spectra of molecules with biradical character [39-43].

6. AN EXAMPLE OF APPLICATION OF P3: 9-METHYLGUANINE

Closed-shell organic molecules are ideal candidates for study with the P3 method for ionization energies. Because of their central position in genetic material as constituents of base pairs, purines and pyrimidines are especially important. The photoelectron spectrum of the purine 9-methylguanine is calculated here as an example of the capabilities of P3 methodology.

Tables 5.2 and 5.3 display vertical ionization energies of the two tautomers (keto and enol) with the lowest energies. The keto form is shown in Fig. 5.1. In the enol form, a proton is transferred from nitrogen 1 to the oxygen atom. P3 ionization energies for both isomers are close to the lowest peak in the photoelectron spectrum (PES) [44].

Table 5.2 Ionization energies (eV) of the keto tautomer of 9-methylguanine.

MO	KT	P3	PES[a]
π_1	8.01	7.98	8.02
π_2	10.57	9.68	9.6
σ_+(N,O)	11.59	9.68	9.6
σ_-(N,O)	11.93	9.91	10.3
π_3	11.64	10.39	10.3
π_4	12.44	11.03	10.86
σ_-N	13.30	11.34	11.32
π_5	14.75	13.14	13.3

[a] Ref. 44.

Table 5.3 Ionization energies (eV) of the enol tautomer of 9-methylguanine.

MO	KT	P3	PES[a]
π_1	8.08	8.05	8.02
π_2	10.18	9.36	9.6
σN	11.26	9.55	9.6
π_3	11.75	10.43	10.3
σN_+	12.38	10.46	10.3
π_4	11.88	10.68	10.86
σN	13.72	11.71	11.32
π_5	14.87	13.31	13.3

[a] Ref. 44.

Dyson orbitals in Figs. 5.2 and 5.3 are distributed similarly in the two tautomers.

A more intense peak at 9.6 eV has several constituent ionization energies corresponding to σ and π holes. Large redistributions of the corresponding Dyson orbitals preserve phase relationships and nodal structure in the π_2 case. The structure of the lowest σ Dyson orbital is preserved between the two tautomers, except for the suppression of the nonbonding lobes on nitrogen 1 or on the oxygen, the positions where the shifting proton may reside. For the feature at 10.3 eV, which has an intensity comparable to the one at 9.6 eV, combinations of σ and π holes also pertain. The order of the π_3 and the second σ hole states changes between the two isomers. There are substantial changes in the

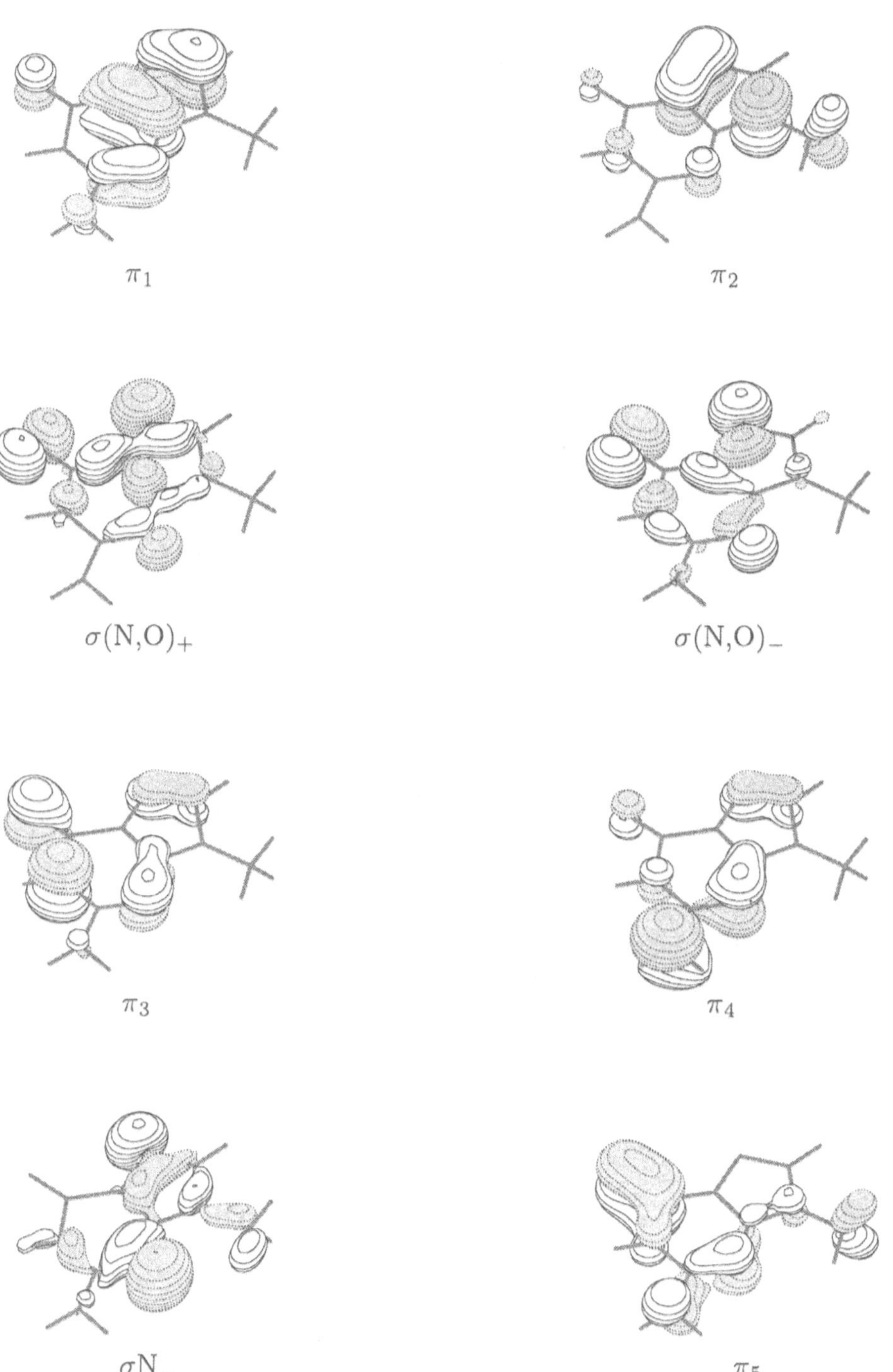

Figure 5.2 Dyson orbitals of the keto form of 9-methylguanine.

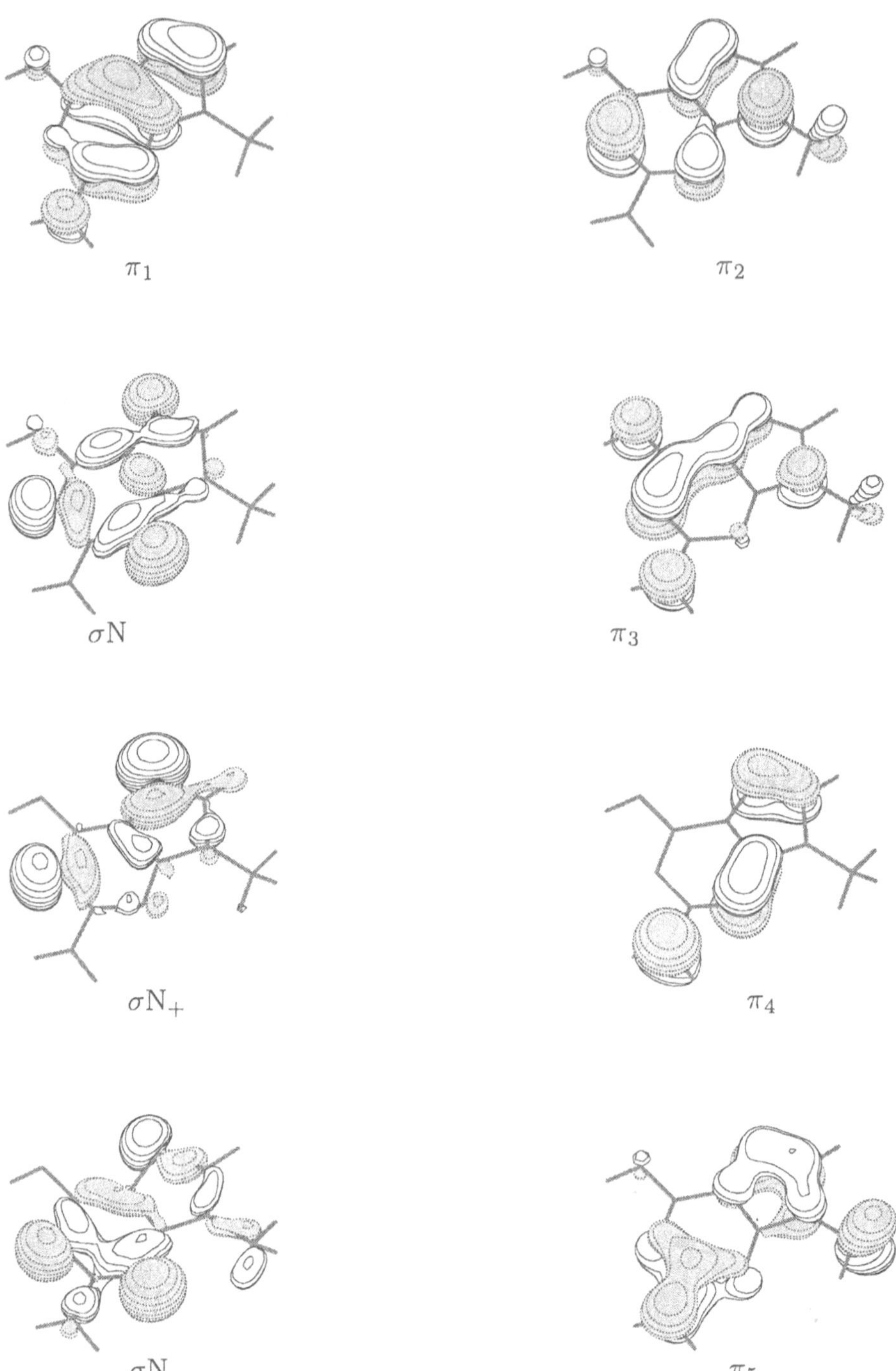

Figure 5.3 Dyson orbitals of the enol form of 9-methylguanine.

Dyson orbitals from one isomer to the other. A higher peak at 10.86 eV, with intensity closer to that of the first one, is assigned to the π_4 hole; the Dyson orbital is approximately conserved after the proton transfer. Note that Koopmans's theorem (KT) results predict the wrong order of states for the enol form. A third σ hole for both isomers is in reasonable agreement with the experimental peak at 11.32 eV. Finally, the feature at 13.3 eV is assigned to the π_5 hole.

The quality of results obtained with the P3/6-311G** model is generally sufficient to assign outer valence photoelectron spectra of typical organic molecules. Only 1s core orbitals are omitted from the self-energy summations of Eq. (3.6). Pole strengths between 0.85 and 0.89 for all states listed here confirm the perturbative arguments on which P3 is based. This kind of calculation can be executed with the standard version of Gaussian 98 [45] by activating certain input keywords [46].

In general, correlation corrections are larger for σ holes than for π holes. It is not unusual for these differential correlation effects to change the predicted order of final states. Heterocyclic organic molecules with nitrogen-centered, nonbonding electrons are not alone in this respect. Organometallics, transition metal complexes, and clusters of metal oxides and metal halides also require this kind of theoretical interpretation.

7. P3 TEST RESULTS

The P3 approximation to the self-energy was applied to the atoms Li through Kr and to neutral and ionic molecular species from the G2 set [47]. For the atoms, a set of 22 representative basis sets was tested. Results for the molecular set were obtained using standard Pople basis sets as described below.

Calculations of ionization energies and electron affinities were performed with a modified development version of Gaussian 99 [48]. Pople and Effective Core Potential (ECP) basis sets are provided in this software [49]. Dunning and Atomic Natural Orbital (ANO) basis sets were obtained from the EMSL Gaussian Basis Set Library [50].

7.1. Atomic Ionization Energies

Atomic calculations are an excellent way to investigate the strengths and weaknesses of a computational procedure's ability to account for electron correlation. Because of the small size of the systems, calculations are sensitive to the theoretical treatment, especially the basis set.

Table 5.4 $4s^n 3d^m$ electron configurations of transition metal atoms.

Atom	Configuration	Multiplicity
Sc	[Ar] $4s^2 3d^1$	2
Ti	[Ar] $4s^2 3d^2$	3
V	[Ar] $4s^2 3d^3$	4
Cr	[Ar] $4s^1 3d^5$	7
Mn	[Ar] $4s^2 3d^5$	6
Fe	[Ar] $4s^2 3d^6$	5
Co	[Ar] $4s^2 3d^7$	4
Ni	[Ar] $4s^2 3d^8$	3
Cu	[Ar] $4s^1 3d^{10}$	2

Fortuitous cancellation of errors is less likely and inherent tendencies of the method under examination may be revealed.

We have performed calculations on the atoms Li through Kr with many basis sets using the P3 method. The results are presented in three groups: alkali and alkaline earth elements, transition metal elements, and p group elements. Electron configurations of transition metal atoms are listed in Table 5.4. For the purposes of this discussion, we have grouped the basis sets into four categories: Pople, Dunning, ECP, and ANO. Data are reported in Tables 5.5, 5.6, and 5.7 for the basis sets that give the best results from each of the four groups. No orbitals were dropped from the summations in the P3 formulae of Eqs. (3.6) and (3.9).

A summary of all calculations is given in Table 5.8. Dunning basis sets are available for p group elements only. Basis set comparisons for other atoms therefore omit this category. Pople bases appear first, followed by the Dunning, ANO, and ECP bases. The mean absolute deviations (MADs) are listed.

P group elements. Molecules with p group elements already have been studied with the P3 approximation and they probably will remain inviting objects of study with this method. Errors obtained for the p group elements (Table 5.5) are somewhat larger than those found for organic molecules. Groups VI and VII are especially problematic.

Results for the other open-shell atoms are encouraging. One would expect the P3 method to be considerably less accurate when an unrestricted Hartree-Fock reference state is used. The lowest MAD for B - Ar obtains with the largest of the Dunning sets examined here, i.e. cc-pVQZ. The 6-311++G(3df,3pd) and well-tempered basis sets (WTBS) are roughly equivalent, with MADs of 0.50 eV and 0.57 eV, respectively.

Table 5.5 Ionization energies and their MADs (eV) computed with the best basis sets for *p* group elements.

Atom	Pople 6-311++G (3df,3pd)	Dunning cc-pVQZ	ANO WTBS	ECP LANL2DZ	Exp.[a]
B	8.10	8.25	7.72	7.68	8.30
C	10.97	11.15	10.87	10.57	11.26
N	14.18	14.38	14.82	13.78	14.54
O	12.82	13.24	14.05	12.35	14.61
F	16.79	17.13	18.21	16.25	17.42
Ne	21.21	21.46	22.67	20.56	21.56
Al	5.64	5.86	5.34	5.43	5.99
Si	7.77	8.03	7.49	7.77	8.15
P	10.14	10.41	9.96	9.99	10.49
S	9.52	10.08	9.34	9.33	10.36
Cl	12.26	12.74	12.41	12.18	13.01
Ar	15.25	15.65	15.48	15.08	15.75
Ga	5.73		5.45	5.37	6.00
Ge	7.66		7.42	7.34	7.88
As	9.73		9.54	9.43	9.82
Se	8.96		8.87	8.59	9.75
Br	11.26		11.29	10.88	11.84
Kr	13.77		13.87	13.41	14.00
MAD	0.50	0.25	0.57	0.82	

[a] Ref. 51.

ECP basis sets performed as expected for these atomic systems, with the Los Alamos double-ζ (LANL2DZ) working best. Its MAD is only 0.82 eV. However, the 31 split-valence ECP of Stevens et al. (CEP-31G) and the Stuttgart-Dresden ECP (SDD) each generated a similar error of 0.87 eV.

It is reasonable to expect P3 calculations with open-shell reference states to be less accurate than their closed-shell counterparts. Unfortunately, there is no obvious correlation between errors and multiplicity.

Errors remain relatively constant for groups III through V, with a sharp increase at group VI. Removal of electrons from β spinorbitals in unrestricted Hartree-Fock reference states is relatively poorly described. Absolute errors for the noble gas elements are significantly lower than

Table 5.6 Ionization Energies and their MADs (eV) computed with the best basis sets for alkali and alkaline earth elements.

Atom	Pople 6-311++G(3df,3pd)	ANO Roos DZ	ECP LANL2DZ	Exp.[a]
Li	5.35	5.36	5.33	5.39
Be	8.81	8.83	8.70	9.32
Na	4.97	4.97	N/A	5.14
Mg	7.24	7.26	N/A	7.64
K	4.22	4.07	3.99	4.34
Ca	5.83	5.84	5.68	6.11
MAD	0.25	0.27	0.37	

[a] Ref. 51.

those for groups VI and VII. Multireference character in O and F militates against the P3 approximation,

Alkali and alkaline earth metals. Results obtained for the group I and group II atoms are encouraging. As Table 5.6 shows, calculations for the alkali atoms are slightly more reliable than those for the alkaline earths. The largest error obtains for the quasidegenerate Be atom. ECP bases provide a convenient alternative to all-electron treatments.

First-row transition metals. These metals present formidable challenges for quantum chemistry. With the energies of the d orbitals being so close to those of the s orbitals for these atoms, the possibility of final states with low pole strengths cannot be ignored. In addition, the middle transition metals are generally difficult to describe with single-determinant methods and require a more advanced approach for a proper description.

In Table 5.7 data are omitted for the Sc and Ti atoms, where pole strengths were well below the acceptable level of 0.80. The remaining results are encouraging. The Roos double-ζ and the ECP SDD sets perform well for Sc through Cr and Ni through Zn. If one ignores the results for Mn through Co, the average absolute error falls sharply to 0.41 eV for the Roos double-ζ basis and to 0.44 eV for the ECP SDD set.

Summary. Despite some noticeable flaws, the performance of the P3 method for atomic calculations is satisfactory. Results for the chalcogens and halogens are somewhat disappointing. In view of the difficulties in describing these atoms with far more complicated methods, this outcome is not surprising. Troublesome results for Mn, Fe, and Co are not

Table 5.7 Ionization energies amd their MADs (eV) computed with the best basis sets for transition metal elements.

Atom	Pople 6-31G(d,p)	ANO Roos DZ	ECP SDD	Exp.[a]
Sc		6.32	6.36	6.54
Ti	6.73		6.52	6.80
V	5.09	6.62	6.66	6.74
Cr	5.76	6.44	6.48	6.76
Mn	5.81	5.58	4.58	7.43
Fe	5.92	6.84	6.75	7.90
Co	6.00	6.89	6.87	7.86
Ni	6.07	7.00	6.92	7.63
Cu	6.14	7.10	6.93	7.72
Zn	8.24	8.87	8.64	9.39
MAD	1.73	0.70	0.81	

[a] Ref. 51.

unexpected. It is perhaps more surprising that the P3 method performs so well for the remainder of the first-row transition metals.

Aside from the results for the individual atoms, some trends in basis set performance may be observed. Pople basis sets produced results that were fairly accurate, especially for alkali and alkaline earth metals. Although the results are much less accurate for the *p* group elements, they are certainly within acceptable error for this simple approximation. The steady decrease in errors observed in the progression from the P3/6-31G to the P3/6-311++G(3df,3pd) level for nontransition elements also attests to the sound design of these basis sets.

The Pople basis sets are perhaps the most efficacious for general applications. Since the integral package in the Gaussian suite of programs is especially efficient with these basis sets, large systems may be tackled routinely in this manner. The Roos double-ζ basis set provides excellent results for all of the elements studied here, including the transition metals. As the Roos triple-ζ basis requires a significant increase in computational cost, this choice is best for smaller systems.

Dunning basis sets have been optimized with atomic configuration interaction calculations and show steady improvement as the basis set quality is increased. The cc-pVQZ set is the most accurate in this category, but its size probably will preclude its use in the larger calculations

Table 5.8 Mean absolute deviations (eV) in the computed ionization energies.

Basis set[a]	Alkali and alkaline earths	*P* group elements	Transition metals
6-31G	0.41	0.92	2.78
6-311G	0.33	0.83	1.99
6-311++G	0.33	0.80	1.78
6-31G(d,p)	0.33	0.66	1.73
6-311++G(3df,3pd)	0.25	0.50	1.93
cc-pVDZ		0.67	
aug-cc-pVDZ		0.51	
cc-pVTZ		0.37	
aug-cc-pVTZ		0.31	
cc-pVQZ		0.25	
WTBS	0.40	0.57	2.67
Wachters			1.47
Ahlrichs	0.46	1.31	1.91
Roos DZ	0.27	0.45	0.70
Roos TZ	0.38	0.13	
CEP-121G		0.91	0.94
CEP-31G		0.87	0.97
CEP-4G		0.67	0.97
LANL2DZ	0.37	0.82	1.91
LANL2MB	0.78	1.48	1.58
SDD	0.63	0.87	0.81
SHC	0.34	0.90	

[a] Refs. 49 and 50.

for which the P3 method is most suited. A steady decline in MAD occurs with increasing size of Dunning basis sets.

Among the ANO basis sets, the Roos double-ζ basis set is clearly preferable. Convergence problems were encountered with Roos triple-ζ basis sets, especially during the pole search in the propagator calculation. Preliminary results are encouraging.

The performance of the P3 method used in conjunction with ECPs is also encouraging. Among the *p* group metals, the CEP-4G set is the most accurate (MAD of 0.67 eV), with the SHC potential of Goddard and Smedley performing best for the alkalis and alkaline earths (MAD of 0.34 eV). The SDD sets succeed in all three cases and produce errors that are competitive with all of the other ECPs.

The overall performance of the P3 method for the atomic systems (see Table 5.8) is encouraging. Transition metals are difficult to describe and the average absolute error for these atoms is generally several times larger than that obtained for the other atoms in the study. However, an unexpected result is the ability of the ECP basis sets to generate errors of less than 1.0 eV for most atoms. Although the errors are relatively large, we anticipate that a combination of ECP and all-electron basis sets will provide an acceptable description of molecules containing these atoms.

7.2. Molecular Species

The G2 set. Calculations of ionization energies and electron affinities for molecules and ions from the G2 set [47] were performed with P3 methods. The diversity of bonding in this set presents a convenient standard for testing the new methodology introduced here, such as electron affinity formulae and procedures for electron binding energies of open-shell systems.

Our implementation computes only vertical ionization energies and electron affinities, but experimental results for the G2 species are adiabatic. To facilitate a direct comparison between the theoretical and experimental results, it is necessary that either the theoretical results be corrected to adiabatic values or that the adiabatic values be related to vertical ones. We have chosen the latter approach and have corrected the experimental results with computational data.

Electron binding energies were calculated in three ways, each involving four steps. The complete procedure is outlined below.

A. Neutral Geometry:

1. HF/6-31G(d) geometry optimization for neutral species with vibrational frequencies to determine zero-point energy.
2. MP2/6-31G(d) geometry optimization for neutral species.
3. Electron propagator calculations
    a) Calculation of electron affinity of the neutral species using the P3 method with 6-311++G(2df,2p) basis set.
    b) Calculation of ionization energy of the neutral species using the P3 method with 6-311G(2df,2p) basis set.
4. Single point energies
    a) Single-point MP2/6-31G(d) calculation for the anionic species at the neutral geometry.
    b) Single-point MP2/6-31G(d) calculation for the cationic species at the neutral geometry.

B. Anion Geometry:

1. HF/6-31G(d) geometry optimization for anionic species with vibrational frequencies to determine the zero-point energy.
2. MP2/6-31G(d) geometry optimization for anionic species.
3. Calculation of ionization energy (electron detachment energy) of the anionic species using P3 method with 6-311++G(2df,2p) basis set.
4. Single-point MP2/6-31G(d) calculation for the neutral species at the anion geometry.

C. Cation Geometry:

1. HF/6-31G(d) geometry optimization for cationic species with vibrational frequencies to determine the zero-point energy.
2. MP2/6-31G(d) geometry optimization for cationic species.
3. Calculation of electron affinities of the cationic species using P3 method with 6-311G(2df,2p) basis set.
4. Single-point MP2/6-31G(d) calculation for the neutral species at the cation geometry.

In cases where experimental data were missing for ionization energies or electron affinities, some steps were omitted. The initial geometries were obtained from the authors of the original G2 study [52].

Transition energies between cations and neutral species were calculated by two procedures. In the first one, the vertical ionization energy of the neutral molecule was determined with the P3 method. These values were compared with experimental adiabatic ionization energies of the neutral molecules, which were adjusted according to

$$\mathrm{IE_N} = \mathrm{IE_{exp}} + \mathrm{Z_N} + \mathrm{R_{N-1}} - \mathrm{Z_{N-1}}\,, \tag{7.1}$$

where $\mathrm{IE_{exp}}$ is the experimental (adiabatic) ionization energy, $\mathrm{Z_N}$ is the zero-point energy (ZPE) of the neutral molecule calculated at the HF/6-31G(d) level, $\mathrm{Z_{N-1}}$ is the ZPE of the cation, and $\mathrm{R_{N-1}}$ is the relaxation energy of the cation between the neutral and cation equilibrium geometries. In other words, each standard of comparison $\mathrm{IE_N}$ is an experimental datum adjusted by calculated zero-point and relaxation energies. In the second procedure, the vertical electron affinity of the cation formed in the first case was computed with the P3 method. These values were compared with experimental, adiabatic electron affinities of the cation (i.e. $\mathrm{IE_{exp}}$), which were adjusted according to

$$\mathrm{EA_{N-1}} = \mathrm{IE_{exp}} + \mathrm{Z_N} - \mathrm{R_N} - \mathrm{Z_{N-1}}\,, \tag{7.2}$$

where EA_{N-1} is the value to which the P3 result is compared.

Transitions between anions and neutral species were also calculated with two procedures. In the first, we calculated the vertical P3 electron affinities of neutral species. The experimental adiabatic electron affinities of the neutral molecules were shifted according to

$$EA_N = EA_{exp} - Z_N - R_{N+1} + Z_{N+1}\,, \tag{7.3}$$

where EA_{exp} is the experimental (adiabatic) electron affinity, Z_N is the ZPE of the neutral molecule, Z_{N+1} is the ZPE of the anion, and R_{N+1} is the relaxation energy of the anion between the neutral and anion equilibrium geometries. In the second procedure, the P3 ionization energy of the anion (that is, the anion's electron detachment energy) was calculated at the anion's geometry. Vertical ionization energies of the anions IE_{N+1} were obtained from experimental adiabatic values, where

$$IE_{N+1} = EA_{exp} + Z_{N+1} + R_N - Z_N\,. \tag{7.4}$$

This sequence of calculations was applied to neutral and ionic molecular species from the G2 test set. Experimental adiabatic electron affinities and ionization energies were taken from Refs. 53 - 74.

Neutral singlets. This class of systems comprises singlet molecules with transitions to doublet cations or anions. Most applications of the P3 method will pertain to such systems.

Electron affinities were calculated as described above with the vertical corrections applied to the experimental results. For molecules with electron affinity data, the overall results are good, with a MAD of just 0.20 eV. The F atom was difficult to describe and this failure is probably related to the low accuracy of the CF_2 calculation, where the error exceeded 1.0 eV.

Similar accuracy obtains when considering the ionization energy (i.e. the electron detachment energy) of the associated doublet anions. Here, MAD is 0.33 eV. For anionic ozone, the error is significantly larger. It is well known that ozone has a great deal of multireference character in its ground state. Ozone therefore is a poor candidate for the P3 method, which relies on the qualitative validity of the single-reference description.

Next we consider ionization energies of neutral singlet states and electron affinities of the associated doublet cations. Here we find that the overall error is slightly larger (MAD of 0.36 eV) than was the case for the electron affinities. Electron affinities of the doublet cations are not treated as well by the P3 method. Here MAD is 0.52 eV.

Neutral doublet states. Electron affinities of neutral doublets where the corresponding anions are singlets, not triplets, are poorly calculated

Table 5.9 MADs for G2 molecules and ions.

Initial geometry	Final state	MAD (eV)	n[a]
^1M	$^2M^-$	0.20	10
$^2M^-$	^1M	0.33	
^1M	$^2M^+$	0.36	35
$^2M^+$	^1M	0.52	
^2M	$^1M^-$	1.11	24
$^1M^-$	^2M	0.20	
^2M	$^1M^+$	0.26	10
$^1M^+$	^2M	0.21	

[a] Number of species [75].

with the P3 method. For these systems, MAD is 1.11 eV. Evaluation of the vertical electron detachment energies of the singlet anions, however, is a more effective approach, for MAD is only 0.20 eV.

For the case of doublet neutrals and associated singlet cations, the ionization energy results are good. Here, the average absolute error is only 0.26 eV. Electron affinities for the singlet cations are also rather accurate. MAD is even less, i.e. 0.21 eV. For energy differences between doublets and triplets, there are not enough systems to establish patterns. These results are highly variable in quality.

Doublet reference states. Some patterns emerge from the calculations with doublet reference states. Table 5.9 presents a summary of all cases involving transitions between singlets and doublets. Ionization energy calculations perform well when a doublet reference state is used. However, electron affinity calculations are advisable only when the doublet reference state is cationic. Even here, it is preferable to reverse the roles of initial and final states by choosing the closed-shell neutral as the reference state in an ionization energy calculation. The P3 method is not suitable for attachment of an electron to a neutral doublet reference state to form a closed-shell anion. It is preferable to choose the anion as the reference state for a P3 calculation of an electron detachment energy. Results for triplets are unpredictable at best.

8. CONCLUSIONS AND PROSPECTUS

P3 calculations with unrestricted Hartree-Fock reference states have been reported here for the first time. In addition, a P3 procedure for electron affinities of closed-shell and open-shell systems has been presented.

Some general trends may be discerned in the P3 results on atoms. Ionization energies involving high-spin states are described well with Pople, Dunning, ANO, and ECP basis sets. Average errors for the *p* block elements are between 0.25 eV for Dunning's quadruple-ζ basis and 0.82 eV for the LANL double-ζ set. For the alkali and alkaline earth metals, the average errors are smaller. Transition metals in the fourth period require use of ANO or SDD sets; the average errors are 0.7 - 0.8 eV. Despite the complex character of electron correlation in $2p$ and $3d$ elements, reasonable results obtain for this simple electron propagator approximation based on an unrestricted Hartree-Fock reference state. For transition metal complexes with high oxidation states and highly electronegative ligands, one may expect the errors for metal-centered holes to be smaller. Results of test calculations using SDD ECPs are especially encouraging for this class of molecules, especially if a closed-shell reference state may be used. Larger errors may be expected for late transition metals, low oxidation states, and relatively electropositive ligands. The P3 method may be used to aid state assignments in photoelectron spectra of organometallics.

Results on molecules and molecular ions display some instructive tendencies. Electron detachment energies from closed-shell reference states (neutral or anionic) are treated well if there is little multiconfigurational character in the initial singlet. Average errors for these cases are about 0.2 - 0.4 eV. The quality of electron attachment energies to closed-shell species is better, especially if the singlet is cationic. Another new class of P3 calculations makes use of unrestricted Hartree-Fock reference states. If the reference state is a doublet, electron detachment energies are treated well, with average errors of 0.2 - 0.3 eV. Electron attachment energies to doublets are satisfactory if the reference state is cationic; caution must be exercised if the doublet reference state is neutral.

Although P3 procedures perform well for a variety of atomic and molecular species, caution is necessary when applying this method to open-shell reference states. Systems with broken symmetry in unrestricted Hartree-Fock orbitals should be avoided. Systems with high multireference character are unlikely to be described well by the P3 or any other diagonal approximation. In such cases, a renormalized electron propagator should be used. In general, P3 will fail when Hartree-Fock theory does not provide a qualitatively acceptable description of the reference state.

The pole strength is a useful diagnostic criterion of problematic cases. Close agreement with experiment in the presence of a pole strength that is less than 0.80 is likely to be the result of a fortuitous cancellation of errors.

The P3 methods for ionization energies and electron affinities provide useful, correlated corrections to canonical Hartree-Fock orbital energies. Their computational demands are modest, especially for electron detachment energies. It is often possible to avoid difficult cases by reversing the labels of initial and final states. Fifth-power arithmetic scaling factors characterize the bottleneck contractions in P3 calculations. Full integral transformations to the Hartree-Fock basis and storage of the largest blocks of integrals may be avoided. In general, P3 calculations may be executed for any molecule where a second-order total energy calculation is feasible. Information on excited final states may be obtained easily. Interpretation of the results in terms of orbitals is facilitated by the diagonal self-energy approximation, where each Dyson orbital is equal to a canonical Hartree-Fock orbital times a scaling factor which is equal to the square root of the corresponding pole strength.

ACKNOWLEDGEMENTS

We thank Dr. Yasuteru Shigeta for informative conversations and assistance in the preparation of this manuscript. This work was supported by the National Science Foundation under grant CHE-9873897 and by the Kansas DEPSCoR program.

REFERENCES

1. J. Linderberg and Y. Öhrn, *Propagators in Quantum Chemistry*, Academic Press, New York (1973).
2. B. T. Pickup and O. Goscinski, *Mol. Phys.* **26** 1013 (1973).
3. L. S. Cederbaum and W. Domcke, *Adv. Chem. Phys.* **36**, 206 (1977).
4. J. Simons, *Annu. Rev. Phys. Chem.* **28**, 1 (1977).
5. J. Simons, *Theor. Chem. Adv. Persp.* **3**, 1 (1979).
6. M. F. Herman, K. F. Freed, and D. L. Yeager, *Adv. Chem. Phys.* **48**, 1 (1981).
7. Y. Öhrn and G. Born, *Adv. Quant. Chem.* **13**, 1 (1981).

8. W. von Niessen, J. Schirmer and L. S. Cederbaum, *Comput. Phys. Rep.* **1**, 57 (1984).
9. J. V. Ortiz in *Computational Chemistry: Reviews of Current Trends*, Vol. 2, J. Leszczynski (Ed.), World Scientific, Singapore (1997) p. 1.
10. J. V. Ortiz in *Conceptual Perspectives in Quantum Chemistry*, Vol. 3, J.-L. Calais and E. Kryachko (Eds.), Kluwer, Dordrecht (1997) p. 465.
11. J. V. Ortiz *Adv. Quantum Chem.* **35**, 33 (1999).
12. J. V. Ortiz, *J. Chem. Phys.* **104**, 7599 (1996).
13. J. V. Ortiz and V. G. Zakrzewski, *J. Chem. Phys.* **105**, 2762 (1996).
14. V. G. Zakrzewski, and J. V. Ortiz, *J. Phys. Chem.* **100**, 13979 (1996).
15. V. G. Zakrzewski, O. Dolgounitcheva, and J. V. Ortiz, *J. Chem. Phys.* **105**, 8748 (1996).
16. V. G. Zakrzewski and J. V. Ortiz, *J. Mol. Struct. (Theochem)* **388**, 351 (1996).
17. J. V. Ortiz, *Int. J. Quantum Chem.* **63**, 291 (1997).
18. M. Enlow, J.V. Ortiz and H.P. Lüthi, *Molecular Physics* **92**, 441 (1997).
19. O. Dolgounitcheva, V. G. Zakrzewski, and J. V. Ortiz, *Int. J. Quantum Chem.* **65**, 463 (1997).
20. O. Dolgounitcheva, V. G. Zakrzewski, and J. V. Ortiz, *J. Phys. Chem. A* **101**, 8554 (1997).
21. V. G. Zakrzewski, O. Dolgounitcheva, and J. V. Ortiz, *J. Chem. Phys.* **107**, 7906 (1997).
22. O. Dolgounitcheva, V. G. Zakrzewski, and J. V. Ortiz, *J. Chem. Phys.* **109**, 87 (1998).
23. O. Dolgounitcheva, V. G. Zakrzewski, and J. V. Ortiz, *Int. J. Quantum Chem.* **70**, 1037 (1998).
24. V. G. Zakrzewski, O. Dolgounitcheva, and J. V. Ortiz, *Int. J. Quantum Chem.* **75**, 607 (1999).
25. O. Dolgounitcheva, V. G. Zakrzewski, and J. V. Ortiz, *Int. J. Quantum Chem.* **80**, 831 (2000).
26. O. Dolgounitcheva, V. G. Zakrzewski, and J. V. Ortiz, *J. Phys. Chem. A* **104**, 10032 (2000).
27. O. Dolgounitcheva, V. G. Zakrzewski, and J. V. Ortiz, *J. Amer. Chem. Soc.* **122**, 12304 (2000).
28. J. M. Herbert and J. V. Ortiz, *J. Phys. Chem.* **A104**, 11786 (2000).
29. H. Hopper, M. Lococo, O. Dolgounitcheva, V. G. Zakrzewski, and J. V. Ortiz, *J. Amer. Chem. Soc.* **122**, 12813 (2000).
30. V. G. Zakrzewski, O. Dolgounitcheva, and J. V. Ortiz, *Int. J. Quant. Chem.* **75**, 607 (1999).
31. V. G. Zakrzewski, J. V. Ortiz, J. A. Nichols, D. Heryadi, D. L. Yeager, and J. T. Golab, *Int. J. Quant. Chem.* **60**, 29 (1996).
32. J. Simons and W. D. Smith, *J. Chem. Phys.* **58**, 4899 (1973).
33. G. D. Purvis and Y. Öhrn, *Chem. Phys. Lett.* **33**, 396 (1975).

34. L. S. Cederbaum, W. Domcke, J. Schirmer, and W. von Niessen, *Adv. Chem. Phys.* **65**, 115 (1986).

35. J. V. Ortiz, *J. Chem. Phys.* **108**, 1008 (1998).

36. O. Dolgounitcheva, V. G. Zakrzewski, and J. V. Ortiz, *J. Chem. Phys.* **114**, 130 (2001).

37. J. T. Golab and D. L. Yeager, *J. Chem. Phys.* **87**, 2925 (1987).

38. D. Heryadi and D. L. Yeager, *J. Chem. Phys.* **114**, 5124 (2001).

39. J. V. Ortiz, *J. Chem. Phys.* **109**, 5741 (1998).

40. J. V. Ortiz, *Int. J. Quant. Chem.* **70**, 651 (1998).

41. J. V. Ortiz, *Chem. Phys. Lett.* **296**, 494 (1998).

42. J. V. Ortiz, *Int. J. Quant. Chem.* **75**, 615 (1999).

43. J. V. Ortiz, *Chem. Phys. Lett.* **297**, 193 (1998).

44. J. Lin, C. Yu, S. Peng, I. Akiyama, K. Li, L. K. Lee, and P. R. Lebreton, *J. Phys. Chem.* **84**, 1006 (1980).

45. GAUSSIAN 98 (Revision A8), M. J. Frisch, G. W. Trucks, H. B. Schlegel, G. E. Scuseria, M. A. Robb, J. R. Cheeseman, V. G. Zakrzewski, J. A. Montgomery, Jr., R. E. Stratmann, J. C. Burant, S. Dapprich, J. M. Millam, A. D. Daniels, K. N. Kudin, M. C. Strain, O. Farkas, J. Tomasi, V. Barone, M. Cossi, R. Cammi, B. Me nnucci, C. Pomelli, C. Adamo, S. Clifford, J. Ochterski, G. A. Petersson, P. Y. Ayala, Q. Cui, K. Morokuma, D. K. Malick, A. D. Rabuck, K. Raghavachari, J. B. Foresman, J. Cioslowski, J. V. Ortiz, B. B. Stefanov, G. Liu, A. Liashenko, P. Piskorz, I. Komaromi, R. Gomperts, R. L. Martin, D. J. Fox, T. Keith, M. A. Al-Laham, C. Y. Peng, A. Nanayakkara, C. Gonzalez, M. Challacombe, P. M. W. Gill, B. Johnson, W. Chen, M. W. Wong, J. L. Andres, M. Head-Gordon, E. S. Replogle, and J. A. Pople, Gaussian, Inc., Pittsburgh PA, 1998.

46. Contact J. V. Ortiz at ortiz@ksu.edu for specific information.

47. L. A. Curtiss, P. C. Redfern, K. Raghavachari, and J. A. Pople, *J. Chem. Phys.* **109**, 42 (1998).

48. GAUSSIAN 99, DEVELOPMENT VERSION (REVISION B.06+), M. J. Frisch, G. W. Trucks, H. B. Schlegel, G. E. Scuseria, M. A. Robb, J. R. Cheeseman, V. G. Zakrzewski, J. A. Montgomery, Jr., R. E. Stratmann, J. C. Burant, S. Dapprich, J. M. Millam, A. D. Daniels, K. N. Kudin, M. C. Strain, O. Farkas, J. Tomasi, V. Barone, B. Mennucci, M. Cossi, C. Adamo, J. Jarmillo, R. Cammi, C. Pomelli, J. Ochterski, G. A. Petersson, P. Y. Ayala, K. Morokuma, D. K. Malick, A. D. Rabuck, K. Raghavachari, J. B. Foresman, J. V. Ortiz, Q. Cui, A. G. Baboul, S. Clifford, J. Cioslowski, B. B. Stefanov, G. Liu, A. Liashenko, P. Piskorz, I. Komaromi, R. Gomperts, R. L. Martin, D. J. Fox, T. Keith, M. A. Al-Laham, C. Y. Peng, A. Nanayakkara, M. Challacombe, P. M. W. Gill, B. Johnson, W. Chen, M. W. Wong, J. L. Andres, C. Gonzalez, M. Head-Gordon, E. S. Replogle, and J. A. Pople, Gaussian, Inc., Pittsburgh PA, 1998.

49. Basis sets (excluding the ANO category) are specified according to the Gaussian 98 keyword used to invoke their use. See A. Frisch and M. J. Frisch, *Gaussian 98 User's Reference* (Gaussian, Inc., Pittsburgh, 1998) pp. 25-28 and references therein for a complete description of these basis sets.

50. Additional basis sets were obtained from the Extensible Computational Chemistry Environment Basis Set Database, June 26, 2000 Version, as developed and

distributed by the Molecular Science Computing Facility, Environmental and Molecular Sciences Laboratory, which is part of the Pacific Northwest Laboratory, P.O. Box 999, Richland, Washington 99352, USA, and is funded by the U.S. Department of Energy. The Pacific Northwest Laboratory is a multiprogram laboratory operated by Battelle Memorial Institute for the U.S. Department of Energy under contract DE-AC06-76RLO 1830.

51. C. E. Moore, *National Standard Reference Data Series 34* (U.S. Government Printing Office, Washington, D.C., 1970).
52. Information on G2 calculations for ionization energies and electron affinities was obtained from the following URL: http://chemistry.anl.gov/compmat/comptherm.htm
53. J. A. Pople, M. Head-Gordon, D. J. Fox, K. Raghavachari, and L. A. Curtiss, *J. Chem. Phys.* **90**, 5622 (1989).
54. L. A. Curtiss, C. Jones, G. W. Trucks, K. Raghavachari, and J. A. Pople, *J. Chem. Phys.* **93**, 2537 (1990).
55. J. Berkowitz, G. B. Ellison, and D. Guttman, *J. Phys. Chem.* **78**, 2744 (1994).
56. H. Hotop and W. C. Lineberger, *J. Phys. Chem. Ref. Data* **14**, 731 (1985).
57. D. W. Arnold, S. E. Bradforth, T. N. Kitsopoulos, and D. M. Neumark, *J. Chem. Phys.* **95**, 8753 (1991).
58. B. Ruscic and J. Berkowitz, *J. Chem. Phys.* **95**, 4033 (1991).
59. K. K. Murray, D. G. Leopold, T. M. Miller, and W. C. Lineberger, *J. Chem. Phys.* **89**, 5442 (1988).
60. S. E. Bradforth, E. H. Kim, D. W. Arnold, and D. M. Neumark, *J. Chem. Phys.* **98**, 800 (1993).
61. K. M. Ervin, J. Ho, and W. C. Lineberger, *J. Phys. Chem.* **92**, 5405 (1988).
62. S. G. Lias, J. E. Bartmess, J. E. Liebman, J. L. Holmes, R. D. Levin, and W. G. Mallard, *J. Phys. Chem. Ref. Data Suppl.* **1**, 17 (1988).
63. M. K. Gilles, M. L. Polak, and W. C. Lineberger, *J. Chem. Phys.* **96**, 8012 (1992).
64. M. K. Gilles, K. M. Ervin, J. Ho, and W. C. Lineberger, *J. Phys. Chem.* **96**, 1130 (1992).
65. H. W. Sarkas, J. H. Hendricks, S. T. Arnold, and K. H. Bowen, *J. Chem. Phys.* **100**, 1884 (1994).
66. J. M. Oakes, L. B. Harding, and G. B. Ellison, *J. Chem. Phys.* **83**, 5400 (1985).
67. *Handbook of Chemistry and Physics*, D. R. Lide (Ed.), CRC, Boca Raton (1996).
68. B. Ruscic and J. Berkowitz, *J. Chem. Phys.* **98**, 2568 (1993).
69. B. Ruscic, E. H. Appelman, and J. Berkowitz, *J. Chem. Phys.* **95**, 7957 (1991).
70. R. J. Lipert and S. D. Coulson, *J. Chem. Phys.* **92**, 3240 (1990).
71. B. Ruscic and J. Berkowitz, *J. Chem. Phys.* **95**, 4378 (1991).
72. J. Berkowitz, E. H. Appelman, and W. A. Chupka, *J. Chem. Phys.* **58**, 1950 (1973).
73. B. Ruscic and J. Berkowitz, *J. Chem. Phys.* **95**, 2407 (1991).
74. B. Ruscic and J. Berkowitz, *J. Chem. Phys.* **95**, 2416 (1991).
75. Singlet electron affinities: H_2CCC, O_3, HNO, LiH, SO_2, S_2O, HCF, CF_2, CH_2S, SiH_2. Singlet ionization energies: HF, H_2O, Si_2H_6, HCl, H_2S, cyclopropene, CS,

CO_2, ClF, CS_2, thiooxirane, Si_2H_2, propyne, SiH_4, Si_2H_4, PH_3, N_2H_2, HOF, NH_3, C_2H_2, BF_3, B_2F_4, BCl_3, B_2H_4, CH_3CHO, CH_3SH, CH_3OH, NCCN, CO, CH_4, CH_3F, CH_3Cl, CF_2, CH_2S, SiH_2. Doublet electron affinities: HCCO, HCO, HOO, NCO, NO_2, OF, CH_3O, NH_2, CH_2CN, CH_2NC, C_2H_3, C_3H_5, CCH, CH_3CH_2O, CH_3S, H_2CCCH, H_2CCHO, CH_3CO, CH_3, CN, OH, PH_2, SH, SiH_3. Doublet ionization energies: Si_2H_5, CH_2SH, C_2H_5, H_2COH, N_2H_3, CH_3, CN, NH_2, PH_2, SiH_3.

Chapter 6

Theoretical Thermochemistry of Radicals

David J. Henry and Leo Radom
Research School of Chemistry, Australian National University, Canberra, ACT 0200, Australia

1. INTRODUCTION

In general, radicals are highly reactive species and can therefore often be difficult to study experimentally [1]. Nevertheless, there is a number of experimental procedures that can be used to determine radical thermochemistry, either directly or indirectly (e.g. through thermochemical cycles) [2]. Berkowitz, Ellison and Gutman [3] have reviewed several of these methods and noted their strengths and limitations. Developments in computer technology mean that ab initio molecular orbital theory [4] now provides a viable alternative source of quantitative gas-phase thermochemical information [5]. However, the theoretical treatment of open-shell systems such as radicals presents its own difficulties [6]. Therefore, the accurate prediction of radical thermochemistry with theoretical procedures poses an interesting challenge.

In this chapter, we look closely at the performance of several ab initio techniques in the prediction of radical thermochemistry with the aim of demonstrating which procedures are best suited in representative situations. We restrict our attention to several areas in which we have had a recent active interest, namely, the determination of radical heats of formation (ΔH_f), bond dissociation energies (BDEs), radical stabilization energies (RSEs), and selected radical reaction barriers and reaction enthalpies. We focus particularly on the results of our recent studies.

J. Cioslowski (ed.), Quantum-Mechanical Prediction of Thermochemical Data, 161–197.

2. THEORETICAL PROCEDURES

There is a wide variety of ab initio techniques available for the study of radical thermochemistry, ranging from quite cheap and approximate methods to much more expensive and accurate approaches. The quality of results yielded by these procedures depends on the size of the basis set used and on the degree of electron correlation included. In practice, it is necessary to strike a balance between the required accuracy and the computational cost that can be afforded.

Commonly-used basis sets include those of Pople and coworkers [4] and Dunning and coworkers [7]. The Pople sets range from small basis sets such as 3-21G, to medium-sized basis sets such as 6-31G(d), to large basis sets such as 6-311+G(3df,2p) or G3large. There are several series of Dunning basis sets including cc-pVnZ, aug-cc-pVnZ (containing diffuse functions), and cc-pCVnZ (containing core-correlation functions). These basis sets increase in size as n goes from D to T, to Q, to 5, to 6, etc..

There is also a hierarchy of electron correlation procedures. The Hartree-Fock (HF) approximation neglects correlation of electrons with antiparallel spins. Increasing levels of accuracy of electron correlation treatment are achieved by Møller-Plesset perturbation theory truncated at the second (MP2), third (MP3), or fourth (MP4) order. Further inclusion of electron correlation is achieved by methods such as quadratic configuration interaction with single, double, and (perturbatively calculated) triple excitations [QCISD(T)], and by the analogous coupled cluster theory [CCSD(T)] [8].

Density functional theory (DFT) [9] is becoming increasingly important in determining chemical properties. Typical methods involve the BLYP functional and the hybrid B3LYP procedure. DFT methods are attractive in that they are often highly cost effective and therefore offer the possibility of application to quite large systems, provided that they are suitably reliable.

Apart from the selection of basis set and correlation procedure, an additional consideration arises in open-shell systems because of the presence of one or more unpaired electrons. This leads to treatments that are referred to as spin-restricted (R), spin-unrestricted (U), and spin-projected (P).

Spin-restricted procedures, signified by an R prefix (e.g. RHF, RMP), constrain the α and β orbitals to be the same. As such, the resulting wavefunctions are eigenfunctions of the spin-squared operator $\langle S^2 \rangle$ that correspond to pure spin states (doublets, triplets, etc). The disadvantage of this approach is that it restricts the flexibility in the electronic description and may result in unrealistic spin localization in radicals.

Spin-unrestricted procedures, designated by the prefix U (e.g. UHF, UMP), treat the α and β electrons independently. This allows more flexibility in accommodating the unpaired electron(s) and, in the case of the Hartree-Fock wavefunction, often leads to a lower-energy description of the electronic structure. However, treating the α and β electrons separately permits the introduction of spin contamination (i.e. mixing of higher spin states) since the wavefunction is no longer an eigenfunction of $\langle S^2 \rangle$. The degree of spin contamination is reflected in the deviation of the $\langle S^2 \rangle$ expectation value from that of a pure spin state (i.e. 0.75 for a doublet, 2.0 for a triplet, etc).

A further alternative is to remove the higher-spin states from the unrestricted wavefunction by means of a spin-projection operator. Spin-projected energies are designated by a P prefix (e.g. PHF, PMP).

It is not clear beforehand which of these alternatives is to be preferred. At the HF and MP levels of theory, the differences between them can be substantial. However, at the QCISD(T) and CCSD(T) levels, it has been found that the differences between the restricted and unrestricted energies are generally small [10, 11].

It has been argued [12] that DFT calculations on open-shell systems should always be performed with spin-unrestricted methods. However, it is still of practical interest to compare the performance of procedures such as UB3LYP and RB3LYP in thermochemical predictions [13].

The ideal calculation would use an infinite basis set and encompass complete incorporation of electron correlation (full configuration interaction). Since this is not feasible in practice, a number of compound methods have been introduced which attempt to approach this limit through additivity and/or extrapolation procedures. Such methods (e.g. G3 [14], CBS-Q [15] and W1′ [16]) make it possible to approximate results with a more complete incorporation of electron correlation and a larger basis set than might be accessible from direct calculations. Table 6.1 presents the principal features of a selection of these methods.

The Gaussian-n (Gn) methods (e.g. G2 [17], G2(MP2, SVP) [18], G3 [14], and G3(MP2) [19]) attempt to approximate the results of a large basis set UQCISD(T) calculation. A defined series of calculations is performed at the UMP2, UMP4, and UQCISD(T) levels of theory with specific basis sets. Additivity approximations are then used to obtain a molecular energy which, when combined with a scaled zero-point vibrational energy (ZPVE) and molecule-independent empirical higher-level correction (HLC), gives the Gn total energy at 0 K for the molecule.

Table 6.1 Principal features of selected compound methods.

Method	Geometry[a]	ZPVE[a]	Energy
G2	UMP2(fu)	UHF	UQCISD(T)/6-311+G(3df,2p)[b]
G2-RAD(QCISD)	UQCISD	UB3LYP	URCCSD(T)/6-311+G(3df,2p)[b]
G2(MP2,SVP)	UMP2(fu)	UHF	UQCISD(T)/6-311+G(3df,2p)[b]
G2(MP2,SVP)-RAD	UB3LYP	UB3LYP	URCCSD(T)/6-311+G(3df,2p)[b]
G3	UMP2(fu)	UHF	UQCISD(T)(fu)/G3large[b]
G3-RAD	UB3LYP	UB3LYP	URCCSD(T)(fu)/G3large[b]
G3(MP2)	UMP2(fu)	UHF	UQCISD(T)/G3MP2large[b]
G3(MP2)-RAD	UB3LYP	UB3LYP	URCCSD(T)/G3MP2large[b]
CBS-Q	UMP2[c]	UHF[c]	UQCISD(T)[d]
CBS-QB3	UB3LYP[e]	UB3LYP[e]	UCCSD(T)[d]
CBS-RAD	UB3LYP	UB3LYP	UCCSD(T)[d]
W1′	UB3LYP[f]	UB3LYP[f]	URCCSD(T)[d]

[a] 6-31G(d) basis set unless noted otherwise.
[b] Estimated.
[c] 6-31G† basis set.
[d] Extrapolated.
[e] 6-311G(2d,d,p) basis set.
[f] cc-pVTZ+1 basis set.

The main feature of the CBS (complete basis set) methods (e.g. CBS-Q [15] and CBS-QB3 [20]) is extrapolation to the complete basis set limit at the UMP2 level. Additional calculations [UMP4 and UQCISD(T) or UCCSD(T)] are performed to estimate higher-order effects. A scaled ZPVE, together with a size-consistent empirical correction and a spin-contamination correction, are added to yield the total CBS energy of the molecule.

Several variations of the Gn and CBS methods have been designed specifically for radicals, and are therefore labeled with the suffix RAD. The Gaussian-n variants include G2-RAD(QCISD) [21], G2(MP2, SVP)-RAD [21], G3-RAD [22], and G3(MP2)-RAD [23]. These procedures are characterized by the use of alternative geometries and scaled zero-point energies, replacement of unrestricted open-shell calculations with restricted open-shell methods, and calculation at the highest correlation level with the URCCSD(T) method instead of UQCISD(T) [24]. CBS-RAD [25] is a variation of the CBS-Q procedure and makes use of a UB3LYP/6-31G(d) geometry and scaled ZPVE while also replacing the UQCISD(T) calculation with UCCSD(T). The principal features of these variants are also included in Table 6.1.

Table 6.2 Comparison of experimental bond lengths (Å) with those calculated[a] with wavefunction-based electronic structure methods.

Radical	Bond	UMP2(fu)	$\langle S^2 \rangle$[b]	RMP2	UQCISD	URCCSD(T)[c]	Exp.[d]
•CH	C–H	1.120	0.756	1.120	1.132	1.122	1.120
•CH_3	C–H	1.078	0.762	1.079	1.084	1.079	1.079
•CH_2CH_3	C–C	1.489	0.763	1.490	1.495	1.492	1.492
•$CHCH_2$	C=C	1.287	0.935	1.317	1.320	1.317	1.316
•CCH	C≡C	1.180	1.187	1.222	1.218	1.215	1.207
	C–H	1.064		1.067	1.069	1.065	1.061
•CO^+	C≡O	1.103	0.941	1.143	1.133	1.123	1.115[e]
•CHO	C=O	1.191	0.762	1.195	1.192	1.183	1.175[e]
	C–H	1.123		1.120	1.125	1.121	1.119[e]
•OCH_3	O–C	1.386	0.758	1.388	1.389	1.376	1.393
•CN	C≡N	1.135	1.127	1.207	1.181	1.179	1.175
•NCH_2	N=C	1.221	0.990	1.259	1.259	1.253	1.247
•CS^+	C≡S	1.459	1.563	1.513	1.502	1.508	1.495[e]
•CF	C–F	1.289	0.759	1.290	1.297	1.278	1.267
•CP	C≡P	1.530	1.545	1.591	1.571	1.578	1.561
•NH_2	N–H	1.028	0.758	1.028	1.034	1.027	1.024
•OH	O–H	0.979	0.755	0.979	0.984	0.971	0.971
•SiH	Si–H	1.526	0.761	1.526	1.539	1.528	1.520
•PH_2	P–H	1.420	0.763	1.419	1.430	1.423	1.429[e]
•SH	S–H	1.344	0.758	1.344	1.353	1.346	1.345
•N_2^+	N≡N	1.147	0.766	1.155	1.133	1.123	1.116
•BO	B=O	1.216	0.800	1.222	1.219	1.214	1.205
•NO	N=O	1.143	0.768	1.177	1.175	1.157	1.151
•ONO	O=N	1.216	0.766	1.223	1.209	1.199	1.197
•O_2^-	O=O	1.380	0.767	1.393	1.358	1.358	1.341
•OF	O–F	1.344	0.766	1.364	1.381	1.357	1.354
•OCl	O–Cl	1.607	0.764	1.613	1.631	1.596	1.570
MAD[f]		0.016		0.015	0.012	0.006	
MD[f]		-0.003		+0.014	+0.012	+0.004	
LD[f]		-0.040		+0.052	+0.030	±0.017	

[a] 6-31G(d) basis set unless noted otherwise; data taken from Ref. 22.
[b] Spin-squared expectation value at UMP2(fu)/6-31G(d).
[c] cc-pVTZ basis set.
[d] Data taken from Ref. 26 unless noted otherwise.
[e] Data taken from Ref. 27.
[f] Excludes the •OCl radical.

A third class of compound methods are the extrapolation-based procedures due to Martin [5], which attempt to approximate infinite-basis-set URCCSD(T) calculations. In the W1′ method [16] calculations are performed at the URCCSD and URCCSD(T) levels of theory with basis sets of systematically increasing size. Separate extrapolations are then performed to determine the SCF, URCCSD valence-correlation, and triple-excitation components of the total atomization energy at

Table 6.3 Comparison of experimental bond lengths (Å) with those calculated with DFT-based electronic structure methods.

Radical	Bond	UB3LYP[a]	UB3LYP[b]	Exp.[c]
•CH	C–H	1.133	1.124	1.120
•CH_3	C–H	1.082	1.078	1.079
•CH_2CH_3	C–C	1.490	1.484	1.492
•$CHCH_2$	C=C	1.310	1.301	1.316
•CCH	C≡C	1.209	1.200	1.207
	C–H	1.068	1.063	1.061
•CO^+	C≡O	1.122	1.110	1.115[d]
•CHO	C=O	1.183	1.173	1.175[d]
	C–H	1.129	1.124	1.119[d]
•OCH_3	O–C	1.369	1.362	1.393
•CN	C≡N	1.174	1.163	1.175
•NCH_2	N=C	1.248	1.238	1.247
•CS^+	C≡S	1.499	1.491	1.495[d]
•CF	C–F	1.288	1.276	1.267
•CP	C≡P	1.565	1.558	1.561
•NH_2	N–H	1.034	1.028	1.024
•OH	O–H	0.983	0.975	0.971
•SiH	Si–H	1.539	1.529	1.520
•PH_2	P–H	1.431	1.426	1.429[d]
•SH	S–H	1.355	1.350	1.345
•N_2^+	N≡N	1.117	1.105	1.116
•BO	B=O	1.209	1.203	1.205
•NO	N=O	1.159	1.146	1.151
•ONO	O=N	1.203	1.192	1.197
•O_2^-	O=O	1.353	1.352	1.341
•OF	O–F	1.354	1.350	1.354
•OCl	O–Cl	1.619	1.593	1.570
MAD[e]		0.008	0.007	
MD[e]		+0.005	-0.003	
LD[e]		-0.024	-0.031	

[a] 6-31G(d) basis set; data taken from Ref. 22.
[b] cc-pVTZ basis set; data taken from Ref. 22.
[c] Data taken from Ref. 26 unless noted otherwise.
[d] Data taken from Ref. 27.
[e] Excludes the •OCl radical.

the basis-set limit. Also included are contributions from core correlation, scaled ZPVE, scalar relativistic effects, and spin-orbit coupling (for atoms only).

3. GEOMETRIES

The accurate determination of thermochemical properties can depend greatly on the quality of the optimized geometry. It is therefore necessary to assess the performance of various procedures for obtaining reliable radical geometries. Tables 6.2 and 6.3 present bond lengths for a selection of radicals [22] optimized at several commonly-used levels of theory and compared with experiment [26, 27]. Also included are mean absolute deviations (MADs), mean deviations (MDs), and largest deviations (LDs) from experiment. A positive sign for an MD or LD indicates an overestimation by a given level of theory.

As can be seen from the mean absolute deviations from experiment, all levels of theory give good overall performance for bond lengths. The poorest result for most of the theoretical procedures is observed for the O–Cl bond length in •OCl, which is overestimated (by 0.023 - 0.061 Å) by all the methods listed in Tables 6.2 and 6.3. This appears to be a consequence of basis set deficiencies, with improved geometries being obtained at all levels of theory with larger basis sets [28]. The •OCl radical has therefore been excluded from the statistical analysis of the results.

The URCCSD(T)/cc-pVTZ level of theory performs the best (Table 6.2), with an MAD of 0.006 Å and an LD of only ±0.017 Å. The positive mean deviation from experiment (+0.004 Å) indicates that this method slightly overestimates most bond lengths. This has also been previously noted by Martin [29].

The DFT-based, computationally inexpensive UB3LYP/6-31G(d) and UB3LYP/cc-pVTZ methods also perform quite well, with MADs of 0.008 Å and 0.007 Å, respectively (Table 6.3). The LDs (-0.024 and -0.031 Å, respectively) for these methods are larger in magnitude than that of URCCSD(T). Martin et al. [30] found for a small set of closed-shell molecules a significant improvement in bond lengths at the UB3LYP level of theory upon going from the cc-pVDZ to the cc-pVTZ basis set. However, for the radicals of Table 6.3, there is very little difference in the performance of the UB3LYP/6-31G(d) and UB3LYP/cc-pVTZ approaches. Bond lengths are slightly overestimated with the 6-31G(d) basis set (MD of +0.005 Å) and slightly underestimated with the cc-pVTZ set (MD of -0.003 Å).

The UQCISD/6-31G(d) level of theory performs marginally less well than its UB3LYP counterpart, with an MAD of 0.012 Å and an LD of +0.030 Å, while slightly overestimating (MD of +0.012 Å) most bond lengths.

Table 6.4 Comparison of experimental bond angles (°) with those calculated[a] with wavefunction-based electronic structure methods.

Radical	Angle	UMP2(fu)	RMP2	UQCISD	URCCSD(T)[b]	Exp.[c]
$\cdot CH_3$	HCH	120.0	120.0	120.0	120.0	120.0
$\cdot CHCH_2$	CCH	136.9	136.1	136.0	136.9	137.3
$\cdot CHO$	HCO	123.4	123.4	124.3	124.4	124.4[d]
$\cdot OCH_3$	OCH	113.6	112.4	112.5	113.9	113.9
$\cdot NCH_2$	HCH	116.7	117.6	116.7	117.6	116.7
$\cdot NH_2$	HNH	103.3	103.0	102.9	102.2	102.9
$\cdot PH_2$	HPH	92.5	92.6	92.1	91.9	91.7[d]
$\cdot ONO$	ONO	133.7	132.4	134.2	134.2	133.9
MAD		0.4	0.9	0.4	0.3	
LD		-1.0	-1.5	-1.4	+0.9	

[a] 6-31G(d) basis set unless noted otherwise; data taken from Ref. 22.
[b] cc-pVTZ basis set.
[c] Data taken from Ref. 26 unless noted otherwise.
[d] Data taken from Ref. 27.

As expected, the UMP2(fu) and RMP2 methods (overall MADs of 0.016 Å and 0.015 Å, respectively) give very similar results for species with minimal spin-contamination. For species displaying significant spin-contamination ($\cdot CHCH_2$, $\cdot CCH$, $\cdot CO^+$, $\cdot CN$, $\cdot NCH_2$, $\cdot CS^+$, and $\cdot CP$), the UMP2 approach generally yields significantly shorter bond lengths than experiment while the RMP2 method often significantly overestimates them, in agreement with previous observations [31]. Large deviations from experiment (> 0.030 Å) are also observed at the UMP2 and RMP2 levels of theory for $\cdot N_2{}^+$ and $\cdot O_2{}^-$. In these two cases, the spin-contamination is small and the UMP2 and RMP2 geometries, although differing significantly from experiment, are quite similar.

As the data in Tables 6.4 and 6.5 indicate, all theoretical levels generally perform well in predicting bond angles at the radical center. MADs range from 0.3° to 0.9° while LDs range from -1.5° to +1.3°.

Overall, among the selected methods, the URCCSD(T)/cc-pVTZ procedure gives the best geometries for the radicals in Tables 6.2 - 6.5. The UB3LYP/6-31G(d) and UB3LYP/cc-pVTZ levels of theory also perform well, and are reasonably economical. The UMP2(fu)/6-31G(d) and RMP2/6-31G(d) approaches generally give acceptable geometries but are not reliable for radicals that display significant spin contamination. This may lead to occasional problems in the calculation of heats of formation for methods that use UMP2 geometries.

Table 6.5 Comparison of experimental bond angles (°) with those calculated with DFT-based electronic structure methods.

Radical	Angle	UB3LYP[a]	UB3LYP[b]	Exp.[c]
$\bullet CH_3$	HCH	120.0	120.0	120.0
$\bullet CHCH_2$	CCH	137.5	138.6	137.3
$\bullet CHO$	HCO	123.6	124.3	124.4[d]
$\bullet OCH_3$	OCH	113.6	113.5	113.9
$\bullet NCH_2$	HCH	116.4	116.7	116.7
$\bullet NH_2$	HNH	102.1	102.7	102.9
$\bullet PH_2$	HPH	91.6	91.8	91.7[d]
$\bullet ONO$	ONO	133.8	134.4	133.9
MAD		0.3	0.3	
LD		-0.8	+1.3	

[a] 6-31G(d) basis set; data taken from Ref. 22.
[b] cc-pVTZ basis set; data taken from Ref. 22.
[c] Data taken from Ref. 26 unless noted otherwise.
[d] Data taken from Ref. 27.

4. HEATS OF FORMATION

The accurate prediction of the heats of formation of molecules has long been one of the main objectives of ab initio molecular orbital procedures [5, 32]. This is particularly important in radical chemistry, where it can be difficult to obtain accurate experimental results. A number of procedures have been used to obtain heats of formation at 0 K ΔH_f^0 from calculated total energies E [33]. We will illustrate them here using the ethyl radical ($\bullet CH_2CH_3$) as an example.

In the atomization approach, the heat of formation of the radical is obtained by combining the calculated energy of the atomization reaction,

$$\bullet CH_2CH_3(g) \rightarrow 2\,C(g) + 5\,H(g)\,, \tag{4.1}$$

with the well-established heats of formation of the gaseous atoms to give

$$\begin{aligned} \Delta H_f^{0,calc}(\bullet CH_2CH_3, g) &= E(\bullet CH_2CH_3) - 2\,E(C) - 5\,E(H) \\ &+ 2\,\Delta H_f^{0,exp}(C, g) + 5\,\Delta H_f^{0,exp}(H, g)\,. \end{aligned} \tag{4.2}$$

Table 6.6 Heats of formation at 0 K (kJ/mol) determined with the Gaussian-2 procedure and its variant.[a]

Radical	$\langle S^2 \rangle$	G2	G2-RAD (QCISD)	Exp.[b]
$\bullet CH_3$	0.762	149.5	152.0	149.8±0.4
$\bullet NH_2$	0.758	191.2	190.1	191.6±1.3[c]
•OH	0.755	37.8	37.2	39.1±0.2
$\bullet SiH_3$	0.754	207.0	207.4	206.7±3.8
$\bullet PH_2$	0.763	141.4	139.0	142.3±2.5
•SH	0.758	144.1	143.5	142.5±3.0
$\bullet N_2^+$	0.766	1506.8	1506.1	1503.3±0.1[c]
•NO	0.768	87.5	90.0	89.8±0.2[d]
•ONO	0.766	32.8	33.5	35.9±0.8[d]
•CN	1.127	445.5	439.5	438.5±4.6
•CCH	1.187	576.6	565.6	561.1±2.9
•CHO	0.762	38.5	39.4	41.3±0.8
$\bullet CHCH_2$	0.935	308.2	307.6	303.8±3.4
$\bullet CH_2CH_3$	0.763	135.5	137.4	131.8±2.1
$\bullet OCH_3$	0.758	27.6	26.2	24.7±3.8
$\bullet SCH_3$	0.758	132.2	131.9	131.4±2.1
$\bullet COCH_3$	0.764	-5.7	-3.4	-3.8±1.3
MAD		3.1	2.1	
MD		+1.6	+0.8	
LD		+15.5	+5.6	

[a] Calculated values from Ref. 22.
[b] Data taken from Ref. 3 unless noted otherwise.
[c] Data taken from Ref. 35.
[d] Data taken from Ref. 26.

In the formation approach, the heat of formation of the radical is obtained from the calculated energy of the formation reaction,

$$2\,\mathrm{C(s)} + \frac{5}{2}\,\mathrm{H_2(g)} \rightarrow \bullet\mathrm{CH_2CH_3(g)}\,, \tag{4.3}$$

as

$$\begin{aligned} \Delta H_f^{0,\mathrm{calc}}(\bullet\mathrm{CH_2CH_3, g}) &= E(\bullet\mathrm{CH_2CH_3}) - 2\,E(\mathrm{C}) \\ &+ 2\,\Delta H_f^{0,\mathrm{exp}}(\mathrm{C, g}) - \frac{5}{2}\,E(\mathrm{H_2})\,. \end{aligned} \tag{4.4}$$

where the energy of the solid state carbon has been replaced by the difference $E(C) - \Delta H_f^{0,\,\mathrm{exp.}}(C, g)$.

Table 6.7 Heats of formation at 0 K (kJ/mol) determined with the Gaussian-3 procedure and its variants.[a]

Radical	$\langle S^2 \rangle$	G3	G3-RAD	G3(MP2)	G3(MP2)-RAD	Exp.[b]
$\bullet CH_3$	0.762	144.8	146.4	145.6	145.0	149.8±0.4
$\bullet NH_2$	0.758	189.2	189.0	189.1	185.6	191.6±1.3[c]
$\bullet OH$	0.755	35.1	34.7	34.7	32.9	39.1±0.2
$\bullet SiH_3$	0.754	207.9	209.9	204.3	203.5	206.7±3.8
$\bullet PH_2$	0.763	140.4	137.7	137.2	132.6	142.3±2.5
$\bullet SH$	0.758	141.0	140.8	138.1	137.0	142.5±3.0
$\bullet N_2^+$	0.766	1510.4	1501.9	1509.3	1505.6	1503.3±0.1[c]
$\bullet NO$	0.768	91.2	89.6	91.7	89.5	89.8±0.2[d]
$\bullet ONO$	0.766	36.9	35.2	41.0	40.1	35.9±0.8[d]
$\bullet CN$	1.127	443.2	437.8	441.9	433.7	438.5±4.6
$\bullet CCH$	1.187	566.6	556.4	564.6	552.9	561.1±2.9
$\bullet CHO$	0.762	40.3	40.2	39.2	38.7	41.3±0.8
$\bullet CHCH_2$	0.935	299.3	298.3	298.4	295.5	303.8±3.4
$\bullet CH_2CH_3$	0.763	130.3	128.6	131.2	129.0	131.8±2.1
$\bullet OCH_3$	0.758	28.2	26.0	29.7	26.1	24.7±3.8
$\bullet SCH_3$	0.758	128.7		126.9	125.2	131.4±2.1
$\bullet COCH_3$	0.764	-4.4	-5.1	-4.1	-4.2	-3.8±1.3
MAD		2.9	2.5	3.6	4.5	
MD		-0.1	-2.0	-0.7	-3.4	
LD		+7.1	-5.5	+6.0	-9.7	

[a] Calculated values from Ref. 22.
[b] Data taken from Ref. 3 unless noted otherwise.
[c] Data taken from Ref. 35.
[d] Data taken from Ref. 26.

It has been found, for G2 calculations on organic molecules in particular, that the atomization method performs somewhat better than the formation method [33].

In the isodesmic approach, the heat of formation of the radical is obtained by combining the calculated energy of an appropriate isodesmic reaction involving the radical, e.g.

$$\bullet CH_2CH_3 + CH_4 \rightarrow CH_3CH_3 + \bullet CH_3 \,, \tag{4.5}$$

with accurate experimental heats of formation of the other species involved in the reaction,

Table 6.8 Heats of formation at 0 K (kJ/mol) determined with the CBS and W1′ procedures.[a]

Radical	CBS-Q	CBS-QB3	CBS-RAD[b]	W1′	Exp.[c]
$\bullet CH_3$	150.1	151.7	153.1	148.3	149.8±0.4
$\bullet NH_2$	193.1	191.5	192.8	188.2	191.6±1.3[c]
$\bullet OH$	37.5	37.9	37.4	36.8	39.1±0.2
$\bullet SiH_3$	203.2	202.0	203.4	200.5	206.7±3.8
$\bullet PH_2$	136.7	135.2	136.6	136.2	142.3±2.5
$\bullet SH$	141.3	140.3	141.3	142.0	142.5±3.0
$\bullet N_2^+$	1509.5	1510.2	1509.6	1510.1	1503.3±0.1[d]
$\bullet NO$	88.3	87.1	87.2	92.8	89.8±0.2[e]
$\bullet ONO$	26.4	30.3	31.4	40.1	35.9±0.8[e]
$\bullet CN$	444.5	442.9	443.4	440.9	438.5±4.6
$\bullet CCH$	570.5	568.9	569.9	563.6	561.1±2.9
$\bullet CHO$	39.5	40.8	41.3	40.9	41.3±0.8
$\bullet CHCH_2$	304.7	304.2	306.1	300.3	303.8±3.4
$\bullet CH_2CH_3$	136.4	135.2	138.3	128.3	131.8±2.1
$\bullet OCH_3$	29.4	25.6	28.5	25.8	24.7±3.8
$\bullet SCH_3$	128.3	127.0	128.7	127.1	131.4±2.1
$\bullet COCH_3$	-4.2	-4.1	-1.7	-5.6	-3.8±1.3
MAD	3.6	3.2	3.6	3.2	
MD	+0.3	-0.2	+1.0	-0.8	
LD	-9.5	+7.8	+8.8	+6.8	

[a] Calculated values from Ref. 22.
[b] UQCISD/6-31G(d) geometry.
[c] Data taken from Ref. 3 unless noted otherwise.
[d] Data taken from Ref. 35.
[e] Data taken from Ref. 26.

$$\begin{aligned}
\Delta H_f^{0,calc}(\bullet CH_2CH_3, g) &= E(\bullet CH_2CH_3) + E(CH_4) \\
&- E(CH_3CH_3) - E(\bullet CH_3) \\
&+ \Delta H_f^{0,exp}(\bullet CH_3, g) + \Delta H_f^{0,exp}(CH_3CH_3, g) \\
&- \Delta H_f^{0,exp}(CH_4, g)\,. \qquad (4.6)
\end{aligned}$$

Because of cancellation of errors in reactions such as (4.5), reasonable results are often obtained, even at quite simple levels of theory. However, it has been found [21, 34] that larger errors may occur with unrestricted methods if there is a significant difference between the degrees of spin contamination for the two radicals in the reaction.

Tables 6.6 - 6.8 present calculated [22] and experimental [3, 26, 35] heats of formation for a selection of small radicals, determined with the atomization approach. As these data show, all theoretical levels give good overall performance (MADs of 2.1 - 4.5 kJ/mol). W1′ is the highest level of theory represented in these tables and indeed performs very well, with an MAD of 3.2 kJ/mol and an LD of +6.8 kJ/mol. The G2-RAD(QCISD) method gives the best statistical performance with an MAD of 2.1 kJ/mol and an LD of +5.6 kJ/mol. However, because of the modest number of comparisons, such differences are only of marginal significance. The G3-RAD approach also performs particularly well with an MAD of 2.5 kJ/mol and an LD of -5.5 kJ/mol. Overall, the G2-RAD(QCISD) method tends to slightly overestimate the heats of formation of the selected radicals (MD of +0.8 kJ/mol) while its G3-RAD counterpart tends to underestimate them (MD of -2.0 kJ/mol). Both of these modified Gn methods offer improved performance over the standard G2 and G3 procedures. However, the G3(MP2)-RAD approach performs less well (MAD of 4.5 kJ/mol) than G3(MP2) (MAD of 3.6 kJ/mol).

The CBS methods all give similar performance, with MADs of 3.2 - 3.6 kJ/mol. The CBS-Q and CBS-RAD variants tend to overestimate the selected radical heats of formation (MDs of +0.3 and +1.0 kJ/mol, respectively), while the CBS-QB3 procedure tends to slightly underestimate them (MD of -0.2 kJ/mol).

With the exception of one G2 case, all levels of theory predict the heats of formation of the selected radicals to within chemical accuracy (i.e. ±10 kJ/mol). The exceptional case is the ethynyl radical, which shows a deviation from experiment of 15.5 kJ/mol at the G2 level of theory. At the G2-RAD(QCISD) level this is reduced to 4.5 kJ/mol. The ethynyl radical exhibits significant spin contamination at the UMP2(fu)/6-31G(d) level and, as noted in the previous section, this leads to a poor geometry (Table 6.2). This is the major cause of the difference between the G2 and G2-RAD(QCISD) values. Similar lowerings of energies (and hence heats of formation) are observed upon going from G3 and G3(MP2) to G3-RAD and G3(MP2)-RAD, respectively. A similar situation is observed for the •CN radical, for which the UMP2(fu)/6-31G(d) geometry is markedly inferior to those computed with the UB3LYP/6-31G(d) and UQCISD/6-31G(d) approaches.

It was also noted in the previous section that the UMP2(fu)/6-31G(d) level of theory performs badly in predicting the geometry of the $\bullet N_2^+$ radical cation while the UB3LYP/6-31G(d) procedure performs quite well. This geometry difference makes a significant contribution to the difference between the G3 and G3-RAD heats of formation for $\bullet N_2^+$.

In summary, all of the methods shown in Tables 6.6 - 6.8 give good overall performance for the prediction of radical heats of formation (MADs of 2.1 - 4.5 kJ/mol). For species displaying significant spin contamination, methods based on a UMP2 reference geometry may give heats of formation for radicals that show larger-than-normal deviations from experiment. The RAD procedures give improved performance in such circumstances.

5. BOND DISSOCIATION ENERGIES

Bond dissociation energies (BDEs) provide a measure of both the reactivity of a compound (with respect to homolytic bond rupture) and the stability of the corresponding radical. There have been many theoretical investigations of BDEs for a wide variety of species [36]. In particular, the C–H BDE for a substituted methane is given by the enthalpy change for the reaction:

$$CH_3X \rightarrow {\bullet}CH_2X + {\bullet}H\,. \tag{5.1}$$

Bond dissociation energies for a selection of substituted methanes, calculated at a range of levels [23], are compared with experimental values [37] in Tables 6.9 and 6.10. Also listed are mean absolute deviations (MADs) and mean deviations (MDs) from experimental values [e.g. MAD(Exp.)] and from CBS-RAD [e.g. MD(CBS-RAD)].

The results in Table 6.9 show that the high-level W1′ procedure generally produces close agreement with experiment, particularly for species with small error bars (< 5 kJ/mol). Several radicals (${\bullet}CH_2SH$, ${\bullet}CH_2C{\equiv}CH$, and ${\bullet}CH_2CHO$) show slightly larger deviations (5.2 - 7.3 kJ/mol), but there is still agreement between theory and experiment to within the given experimental uncertainties. Cyanomethyl and carboxymethyl radicals show the largest deviations between theory and experiment (8.6 and 15.1 kJ/mol, respectively). The mean absolute deviation between W1′ and experiment is only 2.9 kJ/mol for the species with error bars of less than 10 kJ/mol. The W1′ level of theory is therefore considered to be a reliable benchmark level for these systems.

Unfortunately, W1′ is a computationally expensive procedure and therefore not easily accessible for the larger systems listed in Tables 6.9 and 6.10. The CBS-RAD procedure, however, demonstrates close agreement with W1′. For example, the mean absolute deviation between the W1′ and CBS-RAD BDEs is 1.6 kJ/mol while the largest absolute deviation is only 3.3 kJ/mol. Therefore, the CBS-RAD method represents a suitable secondary benchmark level for the assessment of the perfor-

Table 6.9 Comparison of experimental C–H bond dissociation energies at 0 K (kJ/mol) with those calculated[a] with wavefunction-based electronic structure methods.

Radical	RMP2[b]	G3(MP2)-RAD	CBS-RAD	W1′	Exp.[c]
$\bullet CH_3$	413.0	426.4	433.0	432.3	432.2±0.4
$\bullet CH_2NH_2$	367.4	382.1	384.4	383.0	384.4±8.4
$\bullet CH_2OH$	381.2	394.8	398.6	397.0	396.3±1.3
$\bullet CH_2OCH_3$	383.1	395.3	398.7	397.1	395.9
$\bullet CH_2F$	400.6	413.9	419.1	417.5	417.4±4.0
$\bullet CH_2CH_3$	399.4	412.3	417.3	416.4	415.7±1.6
$\bullet CH_2CH_2CH_3$		413.9	419.9		416.6±2.1
$\bullet CH_2CF_3$	421.3	434.1	440.1		439.4±4.6
$\bullet CH_2SH$	378.4	390.3	393.4	391.5	386.3±8.4
$\bullet CH_2Cl$	393.0	405.3	410.0	409.2	411.4±2.3
$\bullet CH_2CH{=}CH_2$	336.0	355.7	359.3	362.0	358.5±4.3
$\bullet CH_2C{\equiv}CH$	362.9	373.8	377.5	378.6	371.3±12.6
$\bullet CH_2C_6H_5$		367.5	377.8		371.1±1.7
$\bullet CH_2CHO$	380.6	391.4	393.0	395.6	389.5±9.2
$\bullet CH_2COOH$	392.7	405.2	408.4	408.6	393.5±12.1
$\bullet CH_2COOCH_3$		404.8	407.9		384.0±12.2
$\bullet CH_2CN$	382.1	394.5	395.8	399.1	390.5±4.4[d]
MAD(Exp.)[e]	15.3	3.6	2.7	2.9	
MD(Exp.)[e]	-15.3	-2.1	+2.5	+2.2	
MAD(CBS-RAD)	16.9	4.3		1.6	
MD(CBS-RAD)	-16.9	-4.3		0.0	

[a] Calculated values from Ref. 23.
[b] RMP2/6-311+G(2df,p)//RMP2/6-31G(d).
[c] Experimental BDEs from Ref. 37 (and references cited therein) unless noted otherwise, and back-corrected to 0 K using theoretical temperature corrections from Ref. 23.
[d] Calculated using experimental heats of formation for $\bullet CH_2CN$ and CH_3CN from Ref. 37.
[e] For species with experimental uncertainties of less than 10 kJ/mol.

mance of other levels of theory in the prediction of BDEs of substituted methanes. The largest deviations between CBS-RAD and experiment occur for $\bullet CH_2COOH$ (14.9 kJ/mol) and $\bullet CH_2COOCH_3$ (23.9 kJ/mol). It has been suggested [23] that experimental re-examination is warranted in these two instances.

Interestingly, with the exception of W1′, all the other levels of theory in Tables 6.9 and 6.10 give BDEs that are smaller than CBS-RAD

Table 6.10 Comparison of experimental C–H bond dissociation energies at 0 K (kJ/mol) with those calculated[a] with the DFT-based electronic structure methods.

Radical	UB3LYP[b]	RB3LYP[c]	Exp.[d]
$•CH_3$	424.6	428.8	432.2±0.4
$•CH_2NH_2$	369.3	373.4	384.4±8.4
$•CH_2OH$	385.3	389.4	396.3±1.3
$•CH_2OCH_3$	384.4	388.4	395.9
$•CH_2F$	406.9	411.1	417.4±4.0
$•CH_2CH_3$	404.4	408.8	415.7±1.6
$•CH_2CH_2CH_3$	407.0	411.3	416.6±2.1
$•CH_2CF_3$	425.7	430.0	439.4±4.6
$•CH_2SH$	380.4	385.4	386.3±8.4
$•CH_2Cl$	398.2	403.1	411.4±2.3
$•CH_2CH{=}CH_2$	345.4	355.5	358.5±4.3
$•CH_2C{\equiv}CH$	358.1	366.6	371.3±12.6
$•CH_2C_6H_5$	357.3	365.4	371.1±1.7
$•CH_2CHO$	379.4	387.0	389.5±9.2
$•CH_2COOH$	394.5	399.5	393.5±12.1
$•CH_2COOCH_3$	394.0	398.9	384.0±12.2
$•CH_2CN$	380.4	388.5	390.5±4.4[e]
MAD(Exp.)[f]	11.2	5.5	
MD(Exp.)[f]	-11.2	-5.5	
MAD(CBS-RAD)	14.1	8.4	
MD(CBS-RAD)	-14.1	-8.4	

[a] Calculated values from Ref. 23.
[b] UB3LYP/6-311+G(3df,2p)//UB3LYP/6-31G(d).
[c] RB3LYP/6-311+G(3df,2p)//RB3LYP/6-31G(d).
[d] Experimental BDEs from Ref. 37 (and references cited therein), unless noted otherwise, and back-corrected to 0 K using theoretical temperature corrections from Ref. 23.
[e] Calculated using experimental heats of formation for $•CH_2CN$ and CH_3CN from Ref. 37.
[f] For species with experimental uncertainties of less than 10 kJ/mol.

values, with the result that the magnitudes of MD(CBS-RAD) and MAD(CBS-RAD) are identical in all these cases.

The G3(MP2)-RAD approach most closely approximates its CBS-RAD counterpart with an MAD of 4.3 kJ/mol. Significantly larger differences (MADs of 8.4 - 16.9 kJ/mol) are observed between CBS-RAD and the RB3LYP, UB3LYP and RMP2 approaches.

Clearly, the W1′ and CBS-RAD methods give quite accurate BDEs for substituted methanes while the G3(MP2)-RAD and RB3LYP methods give acceptable performance. The absolute values of bond dissociation energies obtained from UB3LYP and RMP2 single-point energies are somewhat less satisfactory.

6. RADICAL STABILIZATION ENERGIES

Understanding the stabilizing or destabilizing influence of different substituents on radicals can be particularly important in controlling chemical processes that involve such radicals. Such effects have been studied extensively with electronic structure methods [36].

The radical stabilization energy (RSE) of a substituted methyl radical $\bullet CH_2X$ is generally defined as the difference between the C–H bond dissociation energy in methane and the C–H BDE in the substituted methane CH_3X:

$$RSE(\bullet CH_2X) = BDE(CH_4) - BDE(CH_3X)\,. \qquad (6.1)$$

This quantity is equivalent to the enthalpy change for the isodesmic reaction:

$$\bullet CH_2X + CH_4 \rightarrow CH_3X + \bullet CH_3\,. \qquad (6.2)$$

Radicals with a positive RSE can therefore be considered stabilized relative to $\bullet CH_3$.

It is often assumed that there will be substantial cancellation of errors associated with the calculation of stabilization energies via reactions such as (6.2). However, this is not always the case. In particular, it has recently been shown [21, 34] that stabilization energies calculated for the cyanomethyl and cyanovinyl radicals show large variation with level of theory. For these situations, methods such as UMP2 perform very poorly because errors associated with spin contamination in the reactant and product radicals are very different and do *not* cancel.

Representative radical stabilization energies for the cyanomethyl radical [23, 34, 38] are displayed in Table 6.11. It can be seen that the high-level compound methods all predict RSEs within the narrow range of 31.9 - 37.9 kJ/mol. These values are reasonably close to but slightly lower than the value derived from recent experimental data of 41.7±4.8 kJ/mol (suggesting that the experimental value may be slightly overestimated). On the other hand, the UMP2/6-311+G(3df,2p) approach gives the *wrong sign* for the RSE (-5.7 kJ/mol), i.e. it wrongly predicts a slight destabilizing effect for the cyano substituent. This problem arises because of the artificially high energy calculated within the UMP2 approximation for cyanomethyl radical due to spin contamination. Similar

Table 6.11 Calculated radical stabilization energies at 0 K for the cyanomethyl radical (kJ/mol).[a]

Method[b,c]	RSE	Method[b,c]	RSE
UHF	31.7	G2	32.2
UMP2[d]	-5.7	G2-RAD(QCISD)	32.6
PMP2[e]	22.2	G3	33.1
RMP2	30.8	G3(MP2)-RAD[f]	31.9
UMP4[d,e]	8.8	CBS-Q	37.2
UBLYP[d]	49.9	CBS-QB3	37.9
UB3LYP[f]	44.3	CBS-RAD[f]	37.2
RB3LYP[f]	40.3	CBS-APNO	36.6
UQCISD[d]	33.1	W1′[f]	33.2
UQCISD(T)[d,g]	33.0		
UCCSD(T)[d,g]	31.4		
URCCSD(T)[d,g]	32.9	Exp.[h]	41.7±4.8

[a] Data taken from Ref. 34 unless noted otherwise.
[b] Using 6-311+G(3df,2p) basis set with geometries optimized at the same level of theory but with the 6-31G(d) basis set unless noted otherwise.
[c] UB3LYP/6-31G(d) ZPVEs (scaled by 0.9806) used throughout.
[d] Data taken from Ref. 38.
[e] UMP2/6-31G(d) geometries.
[f] Data taken from Ref. 23.
[g] UQCISD/6-31G(d) geometries.
[h] Calculated using experimental BDEs from Table 6.9.

shortcomings are observed (but to a lesser extent) at the UMP4 (8.8 kJ/mol) and PMP2 (22.2 kJ/mol) levels of theory. Methods such as RMP2, UQCISD, UQCISD(T), UCCSD(T) and URCCSD(T) all perform reasonably well for the RSE of cyanomethyl radical.

RSEs for a broader selection of substituted methyl radicals, as well as MADs and MDs from experiment and CBS-RAD values, are presented in Tables 6.12 and 6.13. We noted in the previous section that our highest-level procedure, namely W1′, gives accurate BDEs, and this observation carries over to the RSEs calculated at this level. The MAD from experiment for the W1′ method is 3.1 kJ/mol. The W1′ RSEs tend to be slightly lower than those determined from experimental data [MD(Exp.) of -2.2 kJ/mol].

At the CBS-RAD level of theory, the MAD from experiment is only 2.4 kJ/mol. Here once again, the CBS-RAD procedure tends to give slightly lower RSEs than experiment [MD(Exp.) of -1.9 kJ/mol].

Table 6.12 Comparison of experimental radical stabilization energies at 0 K (kJ/mol) of substituted methyl radicals with those calculated[a] with wavefunction-based electronic structure methods.

Radical	RMP2[b]	G3(MP2)-RAD	CBS-RAD	W1′	Exp.
$\bullet CH_2NH_2$	45.7	44.2	48.6	49.3	47.8±8.8
$\bullet CH_2OH$	31.8	31.6	34.4	35.3	35.9±1.7
$\bullet CH_2OCH_3$	29.9	31.0	34.3	35.2	36.3
$\bullet CH_2F$	12.4	12.4	13.9	14.8	14.8±5
$\bullet CH_2CH_3$	13.7	14.1	15.7	15.9	16.5±2.0
$\bullet CH_2CH_2CH_3$		12.5	13.1		15.6±2.5
$\bullet CH_2CF_3$	-8.3	-7.7	-7.2		-7.2±5.0
$\bullet CH_2SH$	34.6	36.1	39.6	40.9	45.9±8.8
$\bullet CH_2Cl$	20.0	21.1	23.0	23.1	20.8±2.7
$\bullet CH_2CH{=}CH_2$	77.0	70.7	73.7	70.4	73.7±4.7
$\bullet CH_2C{\equiv}CH$	50.1	52.6	55.5	53.7	60.9±13
$\bullet CH_2C_6H_5$		58.9	55.1		61.1±2.1
$\bullet CH_2CHO$	32.4	34.9	40.0	36.7	42.7±9.6
$\bullet CH_2COOH$	20.3	21.2	24.6	23.8	38.7±12.5
$\bullet CH_2COOCH_3$		21.5	25.1		48.2±12.6
$\bullet CH_2CN$	30.9	31.9	37.2	33.2	41.7±4.8
MAD(Exp.)[c]	4.9	4.1	2.4	3.1	
MD(Exp.)[c]	-4.2	-4.1	-1.9	-2.2	
MAD(CBS-RAD)	3.8	3.0		1.5	
MD(CBS-RAD)	-3.4	-2.5		-0.7	

[a] RSEs calculated using Eq. (6.1) and BDEs from Table 6.9.
[b] RMP2/6-311+G(2df,p)//RMP2/6-31G(d).
[c] For species with experimental uncertainties of less than 10 kJ/mol.

The G3(MP2)-RAD level of theory is found to compare well in accuracy with its CBS-RAD counterpart [MAD(CBS-RAD) of 3.0 kJ/mol]. It can be seen from Table 6.12 that the G3(MP2)-RAD procedure systematically underestimates CBS-RAD stabilization energies [MD(CBS-RAD) of -2.5 kJ/mol]. This appears to be due to the slightly larger deviation in the BDE for methane (6.6 kJ/mol) than for its substituted analogues (ca. 4.1 kJ/mol).

Due to a systematic cancellation of (the quite large) absolute errors in the BDEs, RMP2/6-311+G(2df,p) single-point calculations also perform quite acceptably in predicting RSEs [MAD(CBS-RAD) of 3.8 kJ/mol]. The slightly greater underestimation of the BDE for methane (20.0 kJ/mol) than for the substituted methanes (ca. 16.7 kJ/mol) leads

Table 6.13 Comparison of experimental radical stabilization energies at 0 K (kJ/mol) of substituted methyl radicals with those calculated[a] with DFT-based electronic structure methods.

Radical	UB3LYP[b]	RB3LYP[c]	Exp.
$\bullet CH_2NH_2$	55.3	55.4	47.8±8.8
$\bullet CH_2OH$	39.4	39.4	35.9±1.7
$\bullet CH_2OCH_3$	40.2	40.4	36.3
$\bullet CH_2F$	17.7	17.7	14.8±5
$\bullet CH_2CH_3$	20.2	20.0	16.5±2.0
$\bullet CH_2CH_2CH_3$	17.6	17.5	15.6±2.5
$\bullet CH_2CF_3$	-1.1	-1.2	-7.2±5.0
$\bullet CH_2SH$	44.1	43.4	45.9±8.8
$\bullet CH_2Cl$	26.4	25.7	20.8±2.7
$\bullet CH_2CH{=}CH_2$	79.2	73.3	73.7±4.7
$\bullet CH_2C{\equiv}CH$	66.6	62.2	60.9±13
$\bullet CH_2C_6H_5$	67.3	63.4	61.1±2.1
$\bullet CH_2CHO$	45.2	41.8	42.7±9.6
$\bullet CH_2COOH$	30.1	29.3	38.7±12.5
$\bullet CH_2COOCH_3$	30.6	29.9	48.2±12.6
$\bullet CH_2CN$	44.3	40.3	41.7±4.8
MAD(Exp.)[d]	4.2	3.2	
MD(Exp.)[d]	+4.0	+2.3	
MAD(CBS-RAD)	6.0	4.5	
MD(CBS-RAD)	+6.0	+4.5	

[a] RSEs calculated using Eq. (6.1) and BDEs from Table 6.9.
[b] UB3LYP/6-311+G(3df,2p)//UB3LYP/6-31G(d).
[c] RB3LYP/6-311+G(3df,2p)//RB3LYP/6-31G(d).
[d] For species with experimental uncertainties of less than 10 kJ/mol.

to systematical underestimation of RSEs [MD(CBS-RAD) of -3.4 kJ/mol]. The RB3LYP/6-311+G(3df,2p) level of theory performs slightly less well [MAD(CBS-RAD) of 4.5 kJ/mol] than the RMP2 method for stabilization energies, while the UB3LYP/6-311+G(3df,2p) level shows the largest MAD (6.0 kJ/mol) from CBS-RAD. Interestingly, both RB3LYP and UB3LYP procedures tend to overestimate the CBS-RAD RSEs (MD(CBS-RAD) of +4.5 and +6.0 kJ/mol, respectively). This can be attributed to the fact that both levels underestimate the BDEs for the substituted methanes (ca. 8.7 and 14.4 kJ/mol) to a greater extent than for methane (4.2 and 8.4 kJ/mol, respectively).

7. REACTION BARRIERS

There has been a number of theoretical studies which demonstrate that the accurate prediction of barriers for radical addition reactions is not straightforward [39, 40].

As an illustration of the performance of various levels of theory in determining such barriers, we examine the addition of radicals to alkenes, beginning with methyl radical addition to ethylene,

$$\bullet CH_3 + CH_2 = CH_2 \rightarrow CH_3CH_2CH_2\bullet\,. \tag{7.1}$$

Barriers for reaction (7.1), calculated at a wide variety of levels, are presented in Table 6.14. The theoretical results [41] are compared with the experimental barriers obtained from condensed phase (21.3 kJ/mol) [40, 42] and gas-phase (25.7 kJ/mol) [43] studies, back-corrected for temperature and zero-point energy effects [41, 44].

The first point to note is the extremely large variation in calculated barriers with level of theory that range from 7 kJ/mol (AM1) to 90 kJ/mol (RHF). The UMP2 level of theory gives a barrier of 60 kJ/mol. Higher levels of theory [G2(MP2,SVP), G3, G3(MP2), CBS-RAD, CBS-QB3, and W1′] all give values in the range of 20.5 - 28.6 kJ/mol, which compare favorably with both experimental values. The UB3LYP procedure also performs quite well, giving barriers of 18.3, 25.0, and 25.6 kJ/mol with the 6-31G(d), 6-311+G(d,p), and 6-311+G(3df,2p) basis sets, respectively. Somewhat higher barriers are obtained at the RB3LYP level of theory (23.4, 30.3, and 30.8 kJ/mol, respectively).

The results in Table 6.14 indicate that basis set effects are small, generally less than 7 kJ/mol, but are greater for the DFT-based than for conventional procedures. The barriers tend to increase with basis set size for the B3LYP functional but to decrease with basis set size for the wavefunction-based ab initio methods. The barriers demonstrate much greater sensitivity to the theoretical procedure used.

It is interesting to note that the CBS-RAD and CBS-QB3 levels of theory give barriers (20.5 and 20.7 kJ/mol, respectively) close to the value derived from the condensed-phase experimental measurements, whereas the G2(MP2,SVP), G3(MP2)//B3LYP, G3(MP2)-RAD, G3//B3LYP, and W1′ procedures give results (28.6, 28.3, 27.2, 26.6, and 27.0 kJ/mol, respectively) in accord with the value derived from the gas-phase experimental measurements. It is therefore difficult to provide a definitive assessment of their accuracy at the present time. However, despite these differences, it is important to note that all of the higher-level methods give barriers for the addition of methyl radical to ethylene within approximately a 9 kJ/mol range.

Table 6.14 Calculated barriers (kJ/mol) for the addition of methyl radical to ethylene.[a]

Level of theory	Barrier	Level of theory	Barrier
AM1	7.1	UMP2/6-311+G(3df,2p)[e]	57.5
UHF	39.4	PMP2/6-311+G(3df,2p)[e]	20.9
UMP2	60.1	RMP2/6-311+G(3df,2p)[e]	34.4
UMP4	53.7	UBLYP/6-311+G(3df,2p)[b]	20.7
PMP2	20.1	UB3LYP/6-311+G(3df,2p)[c]	25.6
PMP4	21.5	UB3LYP/6-311+G(d,p)[c]	25.0
RHF	89.7	RB3LYP/6-311+G(d,p)[d]	30.3
RMP2	40.3	RB3LYP/6-311+G(3df,2p)[d]	30.8
RMP4	38.2	UQCISD(T)/6-311+G(d,p)[c]	30.0
UBLYP[b]	13.2	UCCSD(T)/6-311+G(d,p)[c]	30.8
UB3LYP	14.6	URCCSD(T)/6-311+G(d,p)[c]	29.5
UB3LYP[c]	18.3	G2(MP2,SVP)[e]	28.6
RB3LYP[d]	23.4	G3(MP2)//B3LYP[c]	28.3
UQCISD	35.5	G3(MP2)-RAD[c]	27.2
UQCISD(T)	31.7	G3//B3LYP[c]	26.6
UQCISD(T)[e]	33.8	CBS-RAD[c]	20.5
UCCSD(T)[c]	32.7	CBS-RAD[e]	20.9
URCCSD(T)[c]	31.4	CBS-QB3[f]	20.7
URCCSD(T)[e]	33.7	W1′	27.0
		Exp.	21.3[g],25.7[h]

[a] 6-31G(d) basis set, UHF/6-31G(d) geometries unless noted otherwise; data taken from from Ref. 41, calculated in all cases without the inclusion of zero-point vibrational energies.

[b] UBLYP/6-31G(d) geometry.

[c] UB3LYP/6-31G(d) geometry.

[d] RB3LYP/6-31G(d) geometry.

[e] UQCISD/6-31G(d) geometry.

[f] UB3LYP/6-311G(2d,d,p) geometry.

[g] Solution-phase value at 298 K (28.2 kJ/mol) from Refs. 40 and 42, back-corrected to 0 K (+2.2 kJ/mol) and for ZPVE (-9.1 kJ/mol). Corrections from Ref. 41.

[h] Gas-phase value at 403 K (33.1 kJ/mol) from Ref. 43, back-corrected to 0 K (+1.7 kJ/mol) and for ZPVE (-9.1 kJ/mol). Corrections from Ref. 41.

The data compiled in Tables 6.15 and 6.16 indicate how a selection of methods perform in determining reaction barriers for methyl radical additions to a series of substituted alkenes. The experimental values with which comparisons are made in Tables 6.15 - 6.20 come from experiments in solution [40, 42, 45, 46] so there is the possibility of non-negligible solvent effects in some instances.

Table 6.15 Comparison of experimental barriers at 0 K (kJ/mol) for the addition of methyl radical to alkenes $CH_2{=}CXY$ with those calculated[a] with wavefunction-based methods.

X	Y	G3(MP2)-RAD	CBS-RAD	Exp.[b]
H	H	36.2	29.7	30.4
H	Me	34.8	27.6	29.9
H	Et	34.5	26.8	27.5
Me	Me	32.8	25.3	26.8
Me	OMe	36.8		28.2
H	OEt	38.3		28.3
H	OAc	32.4		29.9
H	Cl	29.2	22.2	25.2
H	SiH_3	29.9	23.1	24.8[c]
Me	Cl	27.8	20.4	23.1
Cl	Cl	22.2	16.7	17.4
H	CO_2Me	21.1		17.6
H	CN	21.1	13.8	16.4
H	CHO	22.6	14.6	16.0
Me	CN	20.7		15.3
MAD		5.6	1.7	
MD		+5.6	-1.7	
LD		+10.0	-3.0	
R^2 [d]		0.91	0.98	

[a] Data taken from Ref. 41.

[b] Experimental values from Refs. 40 and 42, back-corrected to 0 K using temperature corrections from Ref. 41.

[c] Experimental barrier for the $SiMe_3$ substituent.

[d] The R^2 values refer to the correlation with experimental barriers.

The UB3LYP/6-31G(d), RB3LYP/6-31G(d) and CBS-RAD procedures perform quite well, with mean absolute deviations from experiment of 2.0, 2.8, and 1.7 kJ/mol, respectively. On the other hand, the UB3LYP/6-311+G(3df,2p), RB3LYP/6-311+G(3df,2p), and G3(MP2)-RAD procedures give larger MADs of 5.2, 9.8, and 5.6 kJ/mol, respectively. All levels of theory display an excellent correlation with experiment (R^2 = 0.90 - 0.98). The UB3LYP/6-31G(d) and CBS-RAD methods tend to give slightly lower barriers than experiment (MD of -1.6 and -1.7 kJ/mol, respectively) while the remaining levels of theory in Tables 6.15 and 6.16 generally give higher barriers than experiment (MD of +2.7 to +9.8 kJ/mol).

Investigation of the addition of substituted methyl radicals to substituted alkenes allows for a broader assessment of the performance of

Table 6.16 Comparison of experimental barriers at 0 K (kJ/mol) for the addition of methyl radical to alkenes $CH_2{=}CXY$ with those calculated with DFT-based electronic structure methods.

X	Y	UB3LYP[a]		RB3LYP[b]		Exp.[c]
		6-31G(d)	large[d]	6-31G(d)	large[d]	
H	H	27.4	34.7	32.5	39.9	30.4
H	Me	27.0	34.5	31.8	39.4	29.9
H	Et	26.9	34.4	31.7	39.4	27.5
Me	Me	26.2	33.2	30.7	38.1	26.8
Me	OMe	29.5	36.7	33.4	40.9	28.2
H	OEt	30.0	37.4	34.2	41.7	28.3
H	OAc	24.6	31.3	28.8	35.8	29.9
H	Cl	22.2	29.3	26.5	33.5	25.2
H	SiH_3	23.1	30.2	27.2	34.6	24.8[e]
Me	Cl	22.2	28.7	26.4	33.3	23.1
Cl	Cl	17.3	23.2	21.0	27.4	17.4
H	CO_2Me	14.5	21.0	21.3	28.5	17.6
H	CN	13.1	19.5	16.3	23.2	16.4
H	CHO	14.3	20.7	17.3	24.2	16.0
Me	CN	14.2	20.5	17.5	24.4	15.3
MAD		2.0	5.2	2.8	9.8	
MD		-1.6	+5.2	+2.7	+9.8	
LD		-5.3	+9.1	+5.9	+13.4	
R^2 [f]		0.90	0.91	0.91	0.91	

[a] UB3LYP/6-31G(d) geometries; data taken from Ref. 41.
[b] RB3LYP/6-31G(d) geometries; data taken from Ref. 41.
[c] Experimental values from Refs. 40 and 42, back-corrected to 0 K using temperature corrections from Ref. 41.
[d] 6-311+G(3df,2p).
[e] Experimental barrier for the $SiMe_3$ substituent.
[f] The R^2 values refer to the correlation with experimental barriers.

the various theoretical procedures, particularly in situations for which there may be significant polar effects. Tables 6.17 and 6.18 list barriers for hydroxymethyl radical additions at a selection of levels of theory.

At the UB3LYP/6-31G(d) and CBS-RAD levels of theory, barriers for the limited number of hydroxymethyl radical additions exhibit larger MADs from experiment (8.5 and 7.3 kJ/mol) than those observed for the methyl radical additions (Tables 6.15 and 6.16). In contrast, barriers obtained at the UB3LYP/6-311+G(3df,2p) and G3(MP2)-RAD levels

Table 6.17 Comparison of experimental barriers at 0 K (kJ/mol) for the addition of hydroxymethyl radical $\bullet CH_2OH$ to alkenes $CH_2{=}CXY$ with those calculated with wavefunction-based electronic structure methods.[a]

X	Y	G3(MP2)-RAD	CBS-RAD	Exp.[b]
H	H	34.1	27.6	34.4
H	Me	34.1	26.8	34.3
H	Cl	30.6	19.2	27.0
H	SiH_3	33.0	18.8	29.3
H	CN	14.7	7.4	14.4
H	CHO	15.8	7.9	11.8
MAD		2.0	7.3	
MD		+1.9	-7.3	
LD		+4.0	-10.5	
R^2		0.96	0.96	

[a] See footnotes to Table 6.15.

[b] Experimental values from Refs. 40 and 45, back-corrected to 0 K using temperature corrections from Ref. 41.

Table 6.18 Comparison of experimental barriers at 0 K (kJ/mol) for the addition of hydroxymethyl radical $\bullet CH_2OH$ to alkenes $CH_2{=}CXY$ with those calculated with the DFT-based electronic structure methods.[a]

X	Y	UB3LYP		RB3LYP		Exp.[b]
		6-31G(d)	large	6-31G(d)	large	
H	H	25.8	35.4	30.7	39.9	34.4
H	Me	26.8	37.2	31.8	42.0	34.3
H	Cl	18.2	27.7	22.5	31.9	27.0
H	SiH_3	18.3	28.1	21.9	31.7	29.3
H	CN	5.2	13.2	7.6	16.0	14.4
H	CHO	5.7	13.6	7.9	16.1	11.8
MAD		8.5	1.5	4.8	4.4	
MD		-8.5	+0.7	-4.8	+4.4	
LD		-11.0	+2.9	-7.4	+7.7	
R^2		0.98	0.98	0.97	0.98	

[a] See footnotes to Table 6.16.

[b] Experimental values from Refs. 40 and 45, back-corrected to 0 K using temperature corrections from Ref. 41.

Table 6.19 Comparison of experimental barriers at 0 K (kJ/mol) for the addition of cyanomethyl radical $\bullet CH_2CN$ to alkenes $CH_2{=}CXY$ with those calculated with wavefunction-based electronic structure methods.[a]

X	Y	G3(MP2)-RAD	CBS-RAD	Exp.[b]
H	H	35.7	28.8	27.5
H	Me	31.2	22.7	23.4
H	Cl	29.4	20.8	23.4
H	SiH_3	31.0	22.9	23.0
H	CN	28.8	18.1	18.2
H	CHO	29.6	18.7	21.6
MAD		8.1	1.3	
MD		+8.1	-0.9	
LD		+10.6	-2.9	
R^2		0.76	0.85	

[a] See footnotes to Table 6.15.

[b] Experimental values from Refs. 40 and 46, back-corrected to 0 K using temperature corrections from Ref. 41.

of theory display good performance, with MADs of 1.5 and 2.0 kJ/mol, respectively. Once again, the UB3LYP/6-31G(d) and CBS-RAD procedures give barriers lower than experiment, while at the G3(MP2)-RAD level barriers are generally higher. Interestingly, the RB3LYP/6-31G(d) method underestimates the barriers for these hydroxymethyl radical additions to a similar extent its RB3LYP/6-311+G(3df,2p) counterpart overestimates them. All levels of theory display a very good correlation with experiment, with R^2 ranging from 0.96 to 0.98.

Calculated barriers for a selection of cyanomethyl radical additions are presented in Tables 6.19 and 6.20. The CBS-RAD method performs particularly well (MAD of 1.3 kJ/mol) for the selected cyanomethyl radical additions. However, the other levels of theory show somewhat larger mean absolute deviations (5.9 - 14.8 kJ/mol). With the exception of CBS-RAD (MD of -0.9 kJ/mol), all levels give higher barriers than those observed experimentally (MD of +5.9 to +14.8 kJ/mol). The correlation with experiment (R^2 = 0.76 - 0.85) is somewhat poorer than that for the methyl and hydroxymethyl radical additions (Tables 6.15 - 6.18).

The overall performance of each of the six levels used to estimate the barriers for radical additions to substituted alkenes (Tables 6.15 - 6.20) is summarized in Table 6.21. The CBS-RAD procedure gives the best overall performance (MAD of 3.2 kJ/mol) and generally underestimates

Table 6.20 Comparison of experimental barriers at 0 K (kJ/mol) for the addition of cyanomethyl radical $\bullet CH_2CN$ to alkenes $CH_2{=}CXY$ with those calculated with DFT-based electronic structure methods.[a]

X	Y	UB3LYP		RB3LYP		Exp.[b]
		6-31G(d)	large	6-31G(d)	large	
H	H	32.3	40.4	34.6	42.6	27.5
H	Me	28.6	36.5	30.1	37.8	23.4
H	Cl	28.5	34.8	30.1	36.2	23.4
H	SiH_3	31.0	38.1	32.6	39.6	23.0
H	CN	25.4	32.5	27.3	34.4	18.2
H	CHO	26.4	33.8	28.0	35.4	21.6
MAD		5.9	13.2	7.6	14.8	
MD		+5.9	+13.2	+7.6	+14.8	
LD		+8.0	+15.1	+9.6	+16.6	
R^2		0.79	0.80	0.79	0.78	

[a] See footnotes to Table 6.16.
[b] Experimental values from Refs. 40 and 46, back-corrected to 0 K using temperature corrections from Ref. 41.

Table 6.21 Overall performance of selected methods in predicting barriers for the addition of methyl, hydroxymethyl and cyanomethyl radicals to substituted alkenes.[a]

	UB3LYP		RB3LYP		G3(MP2)-RAD	CBS-RAD
	6-31G(d)	large[b]	6-31G(d)	large[b]		
MAD	4.3	6.1	4.3	9.8	5.4	3.2
MD	-1.5	+6.0	+2.1	+9.8	+5.3	-2.9
LD	-11.0	+15.1	+9.6	+16.6	+10.6	-10.5
R^2	0.53	0.63	0.64	0.73	0.82	0.78

[a] Using the data of Tables 6.15 - 6.20.
[b] 6-311+G(3df,2p).

the radical addition barriers (MD of -2.9 kJ/mol). The G3(MP2)-RAD method performs less well than its CBS-RAD counterpart (MAD of 5.4 kJ/mol) and tends to overestimate the barriers (MD of +5.3 kJ/mol). The UB3LYP/6-31G(d) and RB3LYP/6-31G(d) results demonstrate the same mean absolute deviation from experiment (4.3 kJ/mol) but whereas the UB3LYP procedure used in conjunction with the small basis set tends to underestimate the radical addition barriers (MD of -1.5 kJ/mol),

Table 6.22 Calculated barriers (kJ/mol) for the ring opening of the cyclopropylcarbinyl radical.[a]

Level of theory	Barrier	Level of theory	Barrier
UMP2/6-311+G(d,p)	66.5	CBS-RAD	32.9
UMP2/6-311+G(3df,2p)	66.6	CBS-Q	32.1
PMP2/6-311+G(d,p)	36.2	G2	37.5
PMP2/6-311+G(3df,2p)	36.3	G2(MP2)	38.2
RMP2/6-311+G(d,p)	43.0	G2(MP2)-RAD	35.5
RMP2/6-311+G(3df,2p)	41.4	G2(MP2,SVP)-RAD	32.7
UB3LYP/6-31G(d)	35.0	G3(MP2)-RAD[b]	31.6
UB3LYP/6-311+G(d,p)	30.3	G3(MP2)-RAD(p)[c]	31.6
UB3LYP/6-311+G(3df,2p)	30.7		
RB3LYP/6-31G(d)[b]	41.0		
RB3LYP/6-311+G(d,p)[b]	35.5		
RB3LYP/6-311+G(3df,2p)[b]	35.8	Exp.[d]	31.2

[a] At UB3LYP/6-31G(d) geometries unless noted otherwise, at 0 K without ZPVE; data taken from Ref. 49 unless noted otherwise.
[b] Data taken from Ref. 38.
[c] UB3LYP/6-31G(d,p) geometry.
[d] Corrected to 0 K with removal of the ZPVE contribution; data taken from Ref. 50.

the RB3LYP procedure overestimates them (MD of +2.1 kJ/mol). A basis-set effect of ca. 7.5 kJ/mol is observed for both the UB3LYP and RB3LYP procedures.

The results presented in Table 6.21 show that the good correlation between calculated and experimental barriers observed for the addition of the *individual* radicals (Tables 6.15 - 6.20) deteriorates significantly when the radicals are examined together. Further work is in progress to try to understand this variation in performance. The largest deviations from experiment are rather larger than desirable (LD ranging from -11.0 to +16.6 kJ/mol).

The ring opening of the cyclopropylcarbinyl radical,

$$\triangleright\!\!-\!\bullet \;\longrightarrow\; \bullet\!\diagdown\!\diagup\!\!\!=\;, \qquad (7.2)$$

has been described as "the most precisely calibrated radical reaction" [47] which makes it ideal for evaluating the performance of theoretical methods. This is one of the fastest unimolecular reactions known [48].

The barriers for this reaction calculated for a range of higher-level procedures (Table 6.22) [49] show that the CBS methods agree well with experiment while their G2 counterparts generally overestimate the

Table 6.23 Calculated barriers (kJ/mol) for the cyclization of the but-3-enyl radical.[a]

Level of theory	Barrier	Level of theory	Barrier
UMP2/6-311+G(d,p)	66.7	CBS-RAD	42.0
UMP2/6-311+G(3df,2p)	65.6	CBS-Q	41.2
PMP2/6-311+G(d,p)	36.3	G2	48.8
PMP2/6-311+G(3df,2p)	35.0	G2(MP2)	49.4
RMP2/6-311+G(d,p)	43.7	G2(MP2)-RAD	45.6
RMP2/6-311+G(3df,2p)	40.6	G2(MP2,SVP)-RAD	44.0
UB3LYP/6-31G(d)	44.5	G3(MP2)-RAD[b]	45.9
UB3LYP/6-311+G(d,p)	45.9	G3(MP2)-RAD(p)[c]	45.9
UB3LYP/6-311+G(3df,2p)	44.8		
RB3LYP/6-31G(d)[b]	50.7		
RB3LYP/6-311+G(d,p)[b]	51.3		
RB3LYP/6-311+G(3df,2p)[b]	50.0	Exp.[d]	38.1,39.2,42.8

[a] At UB3LYP/6-31G(d) geometries unless noted otherwise, and at 0 K without ZPVE; data taken from Ref. 49 unless noted otherwise.
[b] Data taken from Ref. 38.
[c] UB3LYP/6-311G(d,p) geometry.
[d] Corrected to 0 K with removal of the ZPVE. Calculated from recommended experimental enthalpies in Ref. 49.

barrier. The better performance of the CBS methods over the standard G2 approaches has been attributed [49] to the spin-contamination correction that in this instance lowers the barrier by approximately 5 kJ/mol. Of the non-standard methods, the G2(MP2)-RAD procedure offers a slight improvement over its standard G2(MP2) counterpart, while the G2(MP2,SVP)-RAD, G3(MP2)-RAD and G3(MP2)-RAD(p) approaches give results close to the experimental barrier.

The UB3LYP barriers calculated with larger basis sets, namely 6-311+G(d,p) and 6-311+G(3df,2p), are in close agreement with experiment. The RB3LYP barriers are ca. 5.0 kJ/mol higher than their UB3LYP counterparts. The UMP2 approximation performs poorly in estimating the barrier to ring opening. This may be attributed to the significant spin contamination ($\langle S^2 \rangle = 0.95$) observed in the transition structure for this process. The PMP2 and RMP2 methods, while still over-estimating the barrier, offer a significant improvement over UMP2.

Table 6.23 presents calculated barriers for the cyclization of the but-3-enyl radical [i.e. the reverse of reaction (7.2)]. This reaction is an example of an intramolecular radical addition. A number of the features observed in the barriers for the intermolecular radical additions (e.g. methyl radical addition to ethylene, Table 6.14) are also seen here.

Table 6.24 Calculated reaction enthalpies (kJ/mol) for the addition of methyl radical to ethylene.[a]

Level of theory	Enthalpy	Level of theory	Enthalpy
AM1	-157.5	UMP2/6-311+G(3df,2p)[e]	-116.4
UHF	-107.7	PMP2/6-311+G(3df,2p)[e]	-116.9
UMP2	-123.6	RMP2/6-311+G(3df,2p)[e]	-117.0
UMP4	-116.9	UBLYP/6-311+G(3df,2p)[b]	-85.5
PMP2	-123.8	UB3LYP/6-311+G(3df,2p)[c]	-99.0
PMP4	-117.0	RB3LYP/6-311+G(3df,2p)[d]	-97.6
RHF	-107.4	UB3LYP/6-311+G(d,p)[c]	-101.5
RMP2	-124.2	RB3LYP/6-311+G(d,p)[d]	-100.0
RMP4	-117.0	UQCISD(T)/6-311+G(d,p)[c]	-110.5
UBLYP[b]	-107.1	UCCSD(T)/6-311+G(d,p)[c]	-110.4
UB3LYP	-120.1	URCCSD(T)/6-311+G(d,p)[c]	-110.5
UB3LYP[c]	-119.8	G2(MP2,SVP)[e]	-106.5
RB3LYP[d]	-119.1	G3(MP2)//B3LYP[c]	-105.6
UQCISD	-115.6	G3(MP2)-RAD[c]	-105.7
UQCISD(T)	-114.4	G3//B3LYP[c]	-107.4
UQCISD(T)[e]	-112.7	CBS-RAD[c]	-110.0
UCCSD(T)[c]	-112.8	CBS-RAD[e]	-111.5
URCCSD(T)[c]	-112.9	CBS-QB3[f]	-111.2
URCCSD(T)[e]	-112.7	Exp.[g]	-113.1

[a] 6-31G(d) basis set, UHF/6-31G(d) geometries; data taken from Ref. 41. Calculated in all cases without the inclusion of zero-point vibrational energies.
[b] UBLYP/6-31G(d) geometry.
[c] UB3LYP/6-31G(d) geometry.
[d] RB3LYP/6-31G(d) geometry.
[e] UQCISD/6-31G(d) geometry.
[f] UB3LYP/6-311G(2d,d,p) geometry.
[g] Gas-phase value at 298 K (-99 kJ/mol) from Refs. 40 and 43, back-corrected to 0 K (+5.7 kJ/mol) and for ZPVE (-19.8 /mol). Corrections from Ref. 41.

For example, the UMP2 level of theory performs poorly, the RB3LYP approach predicts higher (by ca. 5 - 6 kJ/mol) barriers than its UB3LYP counterpart, and the G3(MP2)-RAD level of theory gives a higher barrier than the CBS-RAD method. One noticeable difference is the absence of a significant basis-set effect in the DFT calculations on the intramolecular addition.

Table 6.25 Comparison of experimental reaction enthalpies at 0 K (kJ/mol) for the addition of methyl radical to alkenes CH_2=CXY with those calculated[a] with the wavefunction-based electronic structure methods.

X	Y	G3(MP2)-RAD	CBS-RAD	Exp.[b]
H	H	-86.3	-91.3	-93.3
H	Me	-86.7	-93.6	-92.2
H	Et	-87.1	-94.6	-94.2
Me	Me	-82.9	-91.6	-93.6
Me	OMe	-77.7		-89.2
H	OEt	-86.7		-95.1
H	OAc	-92.7		-95.9
H	Cl	-98.8	-107.1	-100.8
H	SiH_3	-95.0	-100.6	-95.8[c]
Me	Cl	-96.9	-108.3	-93.6
Cl	Cl	-114.1	-124.6	-119.6
H	CO_2Me	-107.4		-111.5
H	CN	-118.7	-130.4	-133.3
H	CHO	-112.4	-122.9	-116.0
Me	CN	-120.7		-125.0
MAD		6.1	4.7	
MD[d]		+5.7	-3.3	
R^2 [e]		0.90	0.87	

[a] At UB3LYP/6-31G(d) geometries; data taken from Ref. 41.

[b] Experimental estimates from Refs. 40 and 42, back-corrected to 0 K using thermal temperature corrections from Ref. 41.

[c] Experimental value for the $SiMe_3$ substituent.

[d] A negative mean deviation represents an overestimation of the exothermicity while a positive mean deviation represents an underestimation of the exothermicity.

[e] The R^2 values refer to the correlation with experimental reaction enthalpies.

8. REACTION ENTHALPIES

The addition of radicals to alkenes is used to assess the performance of various levels of theory in the prediction of radical reaction enthalpies. Results for the addition of methyl radical to ethylene (Table 6.24) [41] show that the higher-level methods perform well in predicting the reaction enthalpy; values range from -105.6 to -111.5 kJ/mol compared with the corrected experimental value of -113.1 kJ/mol. The AM1 method greatly overestimates the exothermicity while the UB3LYP/6-311+G(3df,2p) level of theory, which performs well for the reaction barrier, significantly underestimates the exothermicity. The RB3LYP values

Table 6.26 Comparison of experimental reaction enthalpies at 0 K (kJ/mol) for the addition of methyl radical to alkenes CH_2=CXY with those calculated[a] with the DFT-based electronic structure methods.

X	Y	UB3LYP/ 6-31G(d)	UB3LYP/ 6-311+G(d,p)	UB3LYP/ 6-311+G(3df,2p)	Exp.[b]
H	H	-100.0	-81.7	-79.3	-93.3
H	Me	-100.8	-82.5	-80.1	-92.2
H	Et	-100.8	-82.8	-80.4	-94.2
Me	Me	-96.5	-78.6	-75.7	-93.6
Me	OMe	-89.6	-72.0	-68.7	-89.2
H	OEt	-99.8	-83.0	-80.0	-95.1
H	OAc	-105.7	-89.1	-86.0	-95.9
H	Cl	-112.3	-93.9	-92.0	-100.8
H	SiH_3	-108.5	-90.6	-88.5	-95.8[c]
Me	Cl	-110.5	-91.8	-89.7	-93.6
Cl	Cl	-127.7	-109.0	-107.5	-119.6
H	CO_2Me	-123.7	-105.8	-103.3	-111.5
H	CN	-136.3	-117.5	-115.4	-133.3
H	CHO	-131.1	-112.7	-110.1	-116.0
Me	CN	-138.1	-120.0	-117.6	-125.0
MAD		8.8	9.2	11.7	
MD[d]		-8.8	+9.2	+11.7	
R^2 [e]		0.91	0.90	0.90	

[a] At UB3LYP/6-31G(d) geometries; data taken from Ref. 41.
[b] Experimental estimates from Refs. 40 and 42, back-corrected to 0 K using thermal temperature corrections from Ref. 41.
[c] Experimental value for the $SiMe_3$ substituent.
[d] A negative mean deviation represents an overestimation of the exothermicity while a positive mean deviation represents an underestimation of the exothermicity.
[e] The R^2 values refer to the correlation with experimental reaction enthalpies.

are very similar to their UB3LYP counterparts for each of the selected basis sets.

Very few directly measured experimental enthalpies are available for methyl radical additions to substituted ethylenes. Reaction enthalpies are therefore normally estimated from other known thermochemical quantities (e.g. C–H BDEs), which often have considerable uncertainties [3], and the derivation generally involves the use of additivity approximations [42, 45]. Therefore, theory may be able to provide more accurate values for these enthalpies. Tables 6.25 and 6.26 present reaction enthalpies determined at several levels of theory and compared with the experimental estimates.

At the UB3LYP level of theory, the MADs range from 8.8 kJ/mol for the 6-31G(d) basis set to 11.7 kJ/mol for the 6-311+G(3df,2p) basis set. Interestingly, the UB3LYP/6-31G(d) level of theory generally overestimates the exothermicity (MD of -8.8 kJ/mol) while the UB3LYP functional used in conjunction with larger basis sets underestimates the experimental exothermicities. The CBS-RAD method gives the best performance with an MAD of 4.7 kJ/mol. The G3(MP2)-RAD method (MAD of 6.1 kJ/mol) performs somewhat better than the UB3LYP functional but also tends to underestimate the exothermicity (MD of 5.7 kJ/mol). In all cases the correlation with experiment is quite good (R^2 spanning the range of 0.87 - 0.91).

9. CONCLUDING REMARKS

The amount of experimental information available regarding the thermochemistry of radicals is limited because of the inherent instability of such species. Therefore, theory has a potentially useful complementary role to play. However, the theoretical determination of radical thermochemistry is not without its own difficulties, and thus a careful assessment of accuracy needs to be carried out before theoretical procedures can be used routinely in this area. Steps in this direction are described in this chapter.

An important general conclusion is that unrestricted procedures such as UMP2 may perform poorly in the case of radicals for which there is significant spin contamination in the underlying UHF wavefunction. Under such circumstances, it is safest to avoid the use of the UHF and UMP2 approximations entirely. The B3LYP procedure appears to be much less sensitive to spin contamination. It is recommended in place of the UHF and UMP2 methods for geometry and frequency predictions in cases where the still more reliable CCSD(T) procedure is not feasible.

High-level compound methods such as Gn, CBS, and W1′ generally perform well in describing radical thermochemistry. However, the reliability of the standard Gn and CBS procedures for radical thermochemistry may generally be improved through modifications (designated RAD) that involve UB3LYP instead of UHF or UMP2 geometries and/or frequencies, RMP2 in place of UMP2 in the additivity steps (of Gn), and CCSD(T) in place of QCISD(T) as the ultimate correlation level. The modified procedures generally perform well in the test cases examined, which include the calculation of heats of formation, bond dissociation energies, radical stabilization energies, and barriers and reaction enthalpies for radical addition reactions.

The B3LYP procedure, while not as reliable for thermochemistry as its higher-level counterparts, is much less expensive computationally and is generally a reasonable cost-effective alternative.

REFERENCES

1. For recent reviews, see: *Energetics of Organic Free Radicals*, J. A. M. Simoes, A. Greenberg, and J. F. Liebman (Eds.), Blackie Academic and Professional, London (1996).

2. For an overview, see: J. A. M. Simoes and M. A. V. R. Da Silva, in *Energetics of Stable Molecules and Reactive Intermediates*, NATO ASI Series C, Vol. 535, M. E. Minas da Piedade (Ed.), Kluwer Academic, Dordecht, The Netherlands (1999), p. 1.

3. J. Berkowitz, G. B. Ellison, and D. Gutman, *J. Phys. Chem.* **98**, 2744 (1994).

4. W. J. Hehre, L. Radom, P. v. R. Schleyer, and J. A. Pople, *Ab Initio Molecular Orbital Theory*, Wiley, New York (1986); F. Jensen, *Introduction to Computational Chemistry*, Wiley, New York (1999).

5. For recent reviews, see for example: *Computational Thermochemistry*, K. K. Irikura and D. J. Frurip (Eds.), ACS Symposium Series, Vol. 677, American Chemical Society, Washington, DC (1998); K. K. Irikura, in *Energetics of Stable Molecules and Reactive Intermediates*, NATO ASI Series C, Vol. 535, M. E. Minas da Piedade (Ed.), Kluwer Academic, Dordecht, The Netherlands (1999), p. 353; J. M. L. Martin, in *Energetics of Stable Molecules and Reactive Intermediates*, NATO ASI Series C, Vol. 535, M. E. Minas da Piedade (Ed.), Kluwer Academic, Dordecht, The Netherlands (1999), p. 373; L. A. Curtiss, P. C. Redfern, and D. J. Frurip, in *Reviews in Computational Chemistry*, Vol. 15, K. B. Lipkowitz and D. B. Boyd (Eds.), Wiley-VCH, New York (2000), p. 147.

6. For a recent review of calculations on open-shell systems, see: T. Bally and W. T. Borden, in *Reviews in Computational Chemistry*, Vol. 13, K. B. Lipkowitz and D. B. Boyd (Eds.), Wiley-VCH, New York (1999), p. 1.

7. For a recent review, see: T. H. Dunning. Jr., K. A. Peterson, and D. E. Woon, in *Encyclopedia of Computational Chemistry*, P. v. R. Schleyer, N. L. Allinger, T. Clark, J. Gasteiger, P. Kollman, H. F. Schaefer III, and P. R. Shreiner (Eds.), Wiley, Chichester (1998), p. 88.

8. In general, in the correlation calculations referred to in this chapter, the core electrons are frozen. In the few cases where they are included in the correlation space, the designation fu (standing for full) is used.

9. See for example: R. G. Parr and W. Yang, *Density Functional Theory of Atoms and Molecules*, Oxford University Press, New York (1989); W. Kohn, A. D. Becke, and R. G. Parr, *J. Phys. Chem.* **100**, 12974 (1996); P. Geerlings, F. De Proft, and W. Langenaeker (Eds.), *Density Functional Theory: A Bridge Between Chemistry and Physics*, VUB Press, Brussels (1999).

10. J. F. Stanton, *J. Chem. Phys.* **101**, 371 (1994).

11. Several of the methods referred to in this chapter use the URCCSD(T) procedure in which a spin-unrestricted CCSD(T) calculation is performed on a high-spin RHF reference wavefunction, as implemented in the MOLPRO program.: H. J. Werner, P. J. Knowles, R. D. Amos, A. Bernhardsson, A. Berning, P. Celani, D. L. Cooper, M. J. O. Deegan, A. J. Dobbyn, F. Eckert, C. Hampel, G. Hetzer, T. Korona, R. Lindh, A. W. Lloyd, S. J. McNicholas, F. R. Manby, W. Meyer, M. E. Mura, A. Nicklass, P. Palmieri, R. Pitzer, G. Rauhut, M. Schtz, H. Stoll,

A. J. Stone, R. Tarroni, and T. Thorsteinsson, MOLPRO 2000.1; University of Birmingham, Birmingham, (1999).

12. J. A. Pople, P. M. W. Gill, and N. C. Handy, *Int. J. Quant. Chem.* **56**, 303 (1995).
13. Indeed, DiLabio et al. have successfully used restricted-open-shell DFT methods to obtain bond dissociation energies: G. A. DiLabio, D. A. Pratt, A. D. LoFaro, and J. S. Wright, *J. Phys. Chem.* **A 103**, 1653 (1999); D. A. Pratt, J. S. Wright, and K. U. Ingold, *J. Am. Chem. Soc.* **121**, 4877 (1999); G. A. DiLabio and D. A. Pratt, *J. Phys. Chem.* **A 104**, 1938 (2000).
14. L. A. Curtiss, K. Raghavachari, P. C. Redfern, V. Rassolov, and J. A. Pople, *J. Chem. Phys.* **109**, 7764 (1998).
15. J. W. Ochterski, G. A. Petersson, and J. J. A. Montgomery, *J. Chem. Phys.* **104**, 2598 (1996).
16. J. M. L. Martin and G. de Oliveira, *J. Chem. Phys.* **111**, 1843 (1999); J. M. L. Martin, *Chem. Phys. Lett.* **310**, 271 (1999).
17. L. A. Curtiss, K. Raghavachari, G. W. Trucks, and J. A. Pople, *J. Chem. Phys.* **94**, 7221 (1991).
18. B. J. Smith and L. Radom, *J. Phys. Chem.* **99**, 6468 (1995); L. A. Curtiss, P. C. Redfern, B. J. Smith, and L. Radom, *J. Chem. Phys.* **104**, 5148 (1996).
19. L. A. Curtiss, P. C. Redfern, K. Raghavachari, V. Rassolov, and J. A. Pople, *J. Chem. Phys.* **110**, 4703 (1999).
20. J. A. Montgomery, Jr., M. J. Frisch, J. W. Ochterski, and G. A. Petersson, *J. Chem. Phys.* **110**, 2822 (1999).
21. C. J. Parkinson, P. M. Mayer, and L. Radom, *Theor. Chem. Acc.* **102**, 92 (1999).
22. D. J. Henry, C. J. Parkinson, and L. Radom, to be published.
23. D. J. Henry, C. J. Parkinson, P. M. Mayer, and L. Radom, *J. Phys. Chem.*, in press.
24. Some of these features are also incorporated in the G2M procedures of Morokuma and co-workers: A. M. Mebel, K. Morokuma, and M. C. Lin, *J. Chem. Phys.* **103**, 7414 (1995).
25. P. M. Mayer, C. J. Parkinson, D. M. Smith, and L. Radom, *J. Chem. Phys.* **108**, 604 (1998); P. M. Mayer, C. J. Parkinson, D. M. Smith, and L. Radom, *J. Chem. Phys.* **108**, 9598 (1998).
26. NIST-JANAF Thermochemical Tables, Fourth Edition, *J. Phys. Chem. Ref. Data*, Monograph No. 9, M. W. Chase, Jr. (Ed.), (1998).
27. Landolt-Börnstein, New Series, *Structure Data of Free Polyatomic Molecules*, K. Kuchitsu (Ed.), Springer, New York (1998-9).
28. For example, the following bond lengths are obtained with the cc-pVQZ basis set: UMP2(fu): 1.552Å, RMP2: 1.557 Å, UB3LYP: 1.584 Å, UQCISD: 1.581 Å, and URCCSD(T): 1.582 Å.
29. J. M. L. Martin, *J. Chem. Phys.* **100**, 8186 (1994).
30. J. M. L. Martin, J. El-Yazal, and J-P. Franois, *Mol. Phys.* **86**, 1437 (1995).
31. D. J. Tozer, N. C. Handy, R. D. Amos, J. A. Pople, R. H. Nobes, Y. Xie, and H. F. Schaefer III, *Mol. Phys.* **79**, 777 (1993).

32. See for example: Y. Fan, in *Encyclopedia of Computational Chemistry*, P. v. R. Schleyer, N. L. Allinger, T. Clark, J. Gasteiger, P. Kollman, H. F. Schaefer III, and P. R. Shreiner (Eds.), Wiley, Chichester (1998), p. 1217; N. L. Allinger, in *Energetics of Stable Molecules and Reactive Intermediates*, NATO ASI Series C, Vol. 535, M. E. Minas da Piedade (Ed.), Kluwer Academic, Dordecht, The Netherlands (1999), p. 417.

33. See for example: A. Nicolaides, A. Rauk, M. N. Glukhovtsev, and L. Radom, *J. Phys. Chem.* **100**, 17460 (1996).

34. C. J. Parkinson, P. M. Mayer, and L. Radom, *J. Chem. Soc., Perkin Trans.* **2**, 2305 (1999).

35. S. G. Lias, J. E. Bartmess, J. F. Liebman, J. L. Holmes, R. D. Levin and W. G. Mallard, *J.Phys. Chem. Ref. Data.* Suppl. 1, 17 (1988); K. P. Huber and G. Herzberg, *Molecular Spectra and Molecular Structure. IV. Constants of Diatomic Molecules*, Van Nostrand Reinhold Co., Princeton (1979).

36. For an extensive set of references, see Ref. 23.

37. *CRC Handbook of Chemistry and Physics*, 80th Edition, D. R. Lide (Ed.), CRC Press, Boca Raton (2000); S. Dóbé, T. Bérces, T. Turányi, F. Márta, J. Grussdorf, F. Temps, and H. G. Wagner, *J. Phys. Chem.* **100**, 19864 (1996); J. A. Seetula, *Phys. Chem. Chem. Phys.* **1**, 4721 (1999); M. S. Robinson, M. L. Polak, V. M. Bierbaum, C. H. DePuy, and W. C. Lineberger, *J. Am. Chem. Soc.* **117**, 6766 (1995); P. G. Wenthold and R. R. Squires, *J. Am. Chem. Soc.* **116**, 11890 (1994); J. L. Holmes, F. P. Lossing, and P. M. Mayer, *J. Am. Chem. Soc.* **113**, 9723 (1991); R. D. Lafleur, B. Szatary, and T. Baer, *J. Phys. Chem.* **A 104**, 1450 (2000).

38. D. J. Henry and L. Radom, to be published.

39. For an extensive set of references, see Ref. 40.

40. H. Fischer and L. Radom, *Angew. Chem., Int. Ed. Engl.*, in press.

41. M. W. Wong and L. Radom, *J. Phys. Chem.* **99**, 8582 (1995); M. W. Wong and L. Radom, *J. Phys. Chem.* **102**, 2237 (1998); D. J. Henry, M. W. Wong, and L. Radom, to be published.

42. T. Zytowski and H. Fischer, *J. Am. Chem. Soc.* **118**, 437 (1996); T. Zytowski and H. Fischer, *J. Am. Chem. Soc.* **119**, 12869 (1997).

43. J. A. Kerr, in *Free Radicals*, J. Kochi (Ed.), Wiley, New York (1972), p. 1; P. M. Holt and J. A. Kerr, *Int. J. Chem. Kinet.* **9**, 185 (1977); D. L. Baulch, C. J. Cobos, R. A. Cox, C. Esser, P. Frank, T. Just, J. A. Kerr, M. J. Pilling, J. Troe, R. W. Walker, and J. Warnatz, *J. Phys. Chem. Ref. Data.* **21**, 411 (1992); D. L. Baulch, C. J. Cobos, R. A. Cox, C. Esser, P. Frank, T. Just, J. A. Kerr, M. J. Pilling, J. Troe, R. W. Walker, and J. Warnatz, *J. Phys. Chem. Ref. Data.* **23**, 847 (1994).

44. See for example: J. I. Steinfeld, J. S. Francisco, and W. L. Hase, *Chemical Kinetics and Dynamics*, Prentice Hall, New Jersey (1989).

45. J. Q. Wu and H. Fischer, *Int. J. Chem. Kinet.* **27**, 167 (1995).

46. J. Q. Wu, I. Beranek, and H. Fischer, *Helv. Chim. Acta.* **78**, 194 (1995).

47. F. N. Martinez, H. B. Schlegel, and M. Newcomb, *J. Org. Chem.* **61**, 8547 (1996).

48. A. L. J. Beckwith and V. Bowry, *J. Am. Chem. Soc.* **116**, 2710 (1994).

49. D. M. Smith, A. Nicolaides, B. T. Golding, and L. Radom, *J. Am. Chem. Soc.* **120**, 10223 (1998).

50. M. Newcomb and A. G. Glenn, *J. Am. Chem. Soc.* **111**, 275 (1989).

Chapter 7

Theoretical Prediction of Bond Dissociation Energies for Transition Metal Compounds and Main Group Complexes with Standard Quantum-Chemical Methods

Nikolaus Fröhlich and Gernot Frenking
Fachbereich Chemie, Philipps-Universität Marburg, Hans-Meerwein Strasse, D-35032, Marburg, Germany

1. INTRODUCTION

In the last decade, quantum-chemical investigations have become an integral part of modern chemical research. The appearance of chemistry as a purely experimental discipline has been changed by the development of electronic structure methods that are now widely used. This change became possible because contemporary quantum-chemical programs provide reliable data and important information about structures and reactivities of molecules and solids that complement results of experimental studies. Theoretical methods are now available for compounds of all elements of the periodic table, including heavy metals, as reliable procedures for the calculation of relativistic effects and efficient treatments of many-electron systems have been developed [1, 2] For transition metal (TM) compounds, accurate calculations of thermodynamic properties are of particularly great usefulness due to the sparsity of experimental data.

For many years, our group has been using quantum-chemical methods to calculate properties of TM compounds. During the last decade, we focused on theoretical analysis of chemical bonds in TM compounds [3-7] and on elucidation of mechanisms of TM-mediated reactions [8].

J. Cioslowski (ed.), Quantum-Mechanical Prediction of Thermochemical Data, 199–233.

Initially, the level of theory that provides accurate geometries and bond energies of TM compounds, yet allows calculations on medium-sized molecules to be performed with reasonable time and CPU resources, had to be determined. Systematic investigations of effective core potentials (ECPs) with different valence basis sets led us to propose a standard level of theory for calculations on TM elements, namely ECPs with valence basis sets of a DZP quality [9, 10]. The small-core ECPs by Hay and Wadt [11] has been chosen, where the original valence basis sets (55/5/N) were decontracted to (441/2111/N-11) with N = 5,4, and 3, for the first-, second-, and third-row TM elements, respectively. The ECPs of the second and third TM rows include scalar relativistic effects while the first-row ECPs are nonrelativistic [11]. For main-group elements, either 6-31G(d) [12-16] all electron basis set or, for the heavier elements, ECPs with equivalent (31/31/1) valence basis sets [17] have been employed. This combination has become our standard basis set II, which is used in a majority of our calculations [18].

Early theoretical work in our group was based on ab initio methods at the Hartree-Fock and correlated levels of theory. Geometries were optimized at the HF/II and MP2/II levels of theory while energies were calculated at the MP2/II and CCSD(T)/II levels. Systematic investigations showed that the HF/II level of theory yields accurate geometries for TM compounds in high oxidation states, whereas its MP2/II counterpart provides reliable geometries of donor-acceptor complexes with metals in low oxidation states [18]. Bond energies calculated at the CCSD(T)/II // HF/II and CCSD(T)/II // MP2/II levels of theory were found to be in excellent agreement with experiment. Bond energies predicted at the MP2/II level are systematically too high but the trends are usually correct. It was shown that the MP2/II level of theory can produce accurate energies when isostructural reactions are employed, leading to error cancellation in the intrinsic deviations of the MP2 results from the CCSD(T) ones [19]. However, it was also found that MP2 calculations on compounds of the first TM row elements, which have partially filled $3d$ shells, frequently yield erroneous results [18].

In the mid-1990s we began to use DFT methods for the calculation of TM compounds. BP86 [20] and B3LYP [21] functionals proved to yield the most accurate results in terms of both geometries and energies. A systematic comparison with our previous ab initio work showed that in many cases BP86 and B3LYP provide bond energies and geometries that are at least as good as those obtained with the MP2 approximation. Compared to the MP2 approach, the DFT methods are significantly more robust for calculations on compounds containing first TM row elements and require much less computer time. This finding allowed us to replace MP2/II with BP86/II and B3LYP/II as our standard levels of theory for calculations on TM compounds. Fortunately, our standard basis set II proved to yield accurate results in conjunction with the DFT methods as well. However, we do not recommend an indiscriminate use of DFT methods, in particular when bond energies are to be calculated without a critical examination of the reliability of the predictions. Like any approximate theoretical method, DFT can produce erroneous results. The drawback of DFT is that it is difficult to predict for which species excessive errors might be anticipated. Unlike ab initio methods, for which the reason of failure can be analyzed in terms of basis set insufficiency and/or inadequate correlation treatment, and where systematic improvement of the theoretical level is possible, DFT methods are much more a black-box tool. The results which are given in this chapter show that sometimes even the trends predicted by the DFT are wrong. For this reason, calculations on a few reference compounds have been carried out at the CCSD(T) level of theory in order to provide reliable data for calibrating the DFT results in cases where no experimental data are available.

In the last decade, more than one hundred theoretical studies of TM compounds have been carried out, addressing questions concerning the geometries, stabilities, reaction mechanisms, chemical and physical properties, and the bonding situations in a wide variety of molecules [1-7, 18, 22]. Many publications reported bond dissociation energies (BDEs) of TM–ligand bonds. As mentioned above, very few reliable experimental BDEs of TM compounds are available in the literature [18, 22]. Because the theoretical predictions for BDEs at the CCSD(T)/II level of theory were found to be very accurate, they have been used to calibrate the data obtained at the MP2/II, BP86/II and B3LYP/II levels. Therefore, these BDEs can be regarded as a valuable source of thermodynamic data on TM compounds. Since the theoretically predicted BDEs have not been previously summarized, in this chapter we provide a compilation of the calculated data which should be helpful as a reference. Although the majority of our work focused on TM compounds, we also calculated BDEs of main group complexes of the group-13 Lewis acids EX_3 (E = B - In; X = H, F, Cl) and BeO with various Lewis bases. A list of these theoretically predicted values is included in this work.

As mentioned above, most calculations were carried out with our standard basis set II. The compilation of results in the present form makes it possible to compare a great variety of fairly large compounds studied at the same level of theory. We are well aware of the fact that the apparently high accuracy of the theoretical values is partly due to a fortuitous error cancellation. Thus, when it was necessary to obtain more accurate data for a given project, we employed the Stuttgart ECPs that are associated with larger basis sets and include scalar relativistic effects also for first TM row elements [17, 23]. This is particularly important for copper, because relativistic effects have a significant impact on the calculated geometries and BDEs of its compounds [24-28].

We want to emphasize that other groups have also published accurate quantum-chemical calculations of TM–ligand bond energies. The fact that we do not include their results in the present compilation does not imply that we consider them less accurate than ours. Theoretical studies that predict BDEs of TM compounds at a much higher level of theory than those employed in our work have been published but the molecules studied have been smaller than the species listed in this chapter. Here we intend to compare data calculated at a level of theory that can be applied to larger molecules and that has been used for a large number of different classes of compounds. A uniform level of theory makes it possible to directly compare the strengths of bonds between different ligands and different metals. We are planning to create a data bank which contains all of the calculated BDEs. This continuously updated data bank will be available online.

The presentation of the theoretical BDEs is organized as follows. We list the calculated D_e values and their zero-point energy corrected counterparts (denoted by D_0) for TM compounds that belong to different classes of molecules. This leads in some cases to double presentation, as some species belong to more than one class. However, we believe that the ordering chosen here facilitates comparison between compounds as well as between methods in a balanced way.

Most calculations were carried out with the Gaussian 94 [29], Gaussian 98 [30], Turbomole [31], ACES II [32] and MOLPRO [33] program packages. We also used the DFT program ADF [34], which is different from the other electronic structure software as it uses Slater-type orbitals rather than Gaussian functions. The calculations of the TM compounds with ADF were carried out at the BP86 level of theory with a numerical TZP-quality basis set. Relativistic effects were calculated either with the Pauli Hamiltonian [35] or, recently, with the more reliable ZORA approximation [36]. Details about the computational procedures can be found in the original publications.

Table 7.1 Calculated first BDEs D_e (D_0) (kcal/mol) of neutral homoleptic carbonyl complexes.[a]

Molecule	MP2/II	BP86/II	BP86/TZP	CCSD(T)/II[b]	Exp.
$Ni(CO)_4$	49.1 (47.8)	28.4 (26.3)	29.1 (27.0)	23.6 (22.3)	25±2[c]
$Pd(CO)_4$	14.8 (14.5)	14.2 (12.9)	13.0 (11.7)	7.8 (7.5)	
$Pt(CO)_4$	21.6 (20.4)	15.7 (14.2)	14.9 (13.5)	12.1 (10.9)	
$Fe(CO)_5$	66.1 (63.6)	48.7 (45.9)	46.8 (44.0)	46.9 (44.4)	41±2[d]
$Ru(CO)_5$	41.2 (39.1)			30.9 (28.8)	27.6±0.4[e]
$Os(CO)_5$	52.7 (50.1)			42.9 (40.3)	30.6±0.3[f]
$Cr(CO)_6$	58.0 (55.4)			45.8 (43.2)	36.8±2[d]
$Mo(CO)_6$	46.1 (43.9)			40.4 (38.2)	40.5±2[d]
$W(CO)_6$	54.9 (52.6)	49.4 (47.3)[g]	46.0 (43.9)	48.0 (45.7)	46.0±2[d]

[a] BDEs taken from Refs. 37-42.
[b] At MP2/II optimized geometries.
[c] Ref. 43.
[d] Ref. 44.
[e] Ref. 45.
[f] Ref. 46.
[g] The corresponding B3LYP/II BDE equals 45.9 (43.8) kcal/mol.

2. HOMOLEPTIC CARBONYL COMPLEXES

The number of TM carbonyl complexes for which experimental BDEs are known is relatively large. Theoretical studies of neutral 18-valence electron carbonyl complexes of groups 6, 8 and 10 have been reported [37-42]. Table 7.1 lists the calculated and experimental values of D_e and D_0.

In most cases, the computed values of D_0 are in a very good agreement with experiment. Although the calculated BDEs of $Os(CO)_5$ and $Cr(CO)_6$ are higher than their experimental counterparts, a closer study of the latter suggests that the reported values may be too low. Inspection of Table 7.1 leads to the conclusion that the performance of the BP86/II and B3LYP/II levels of theory is very good as the DFT BDEs are close to their CCSD(T) counterparts. The MP2/II data are always too high but the trends in the calculated BDEs are correct. Note that the MP2/II level of theory yields particularly high BDEs for $Ni(CO)_4$ and $Fe(CO)_5$. These erroneously high values are representative of the problems that are often encountered when first TM row compounds are treated at the MP2 level of theory.

Table 7.2 Calculated first BDEs D_e (D_0) (kcal/mol) of charged carbonyl complexes.[a,b]

Molecule	MP2/II[c]	BP86/II[c]	BP86/ TZP[c]	B3LYP/ II[c]	CCSD(T)/ II[c,d,e]	Exp.[e]
$Hf(CO)_6^{2-}$		54.9 (53.2)	50.8	51.4 (49.6)		
$Ta(CO)_6^-$	53.1 (51.2)	50.9 (49.1)	48.3	47.8 (46.0)	47.9 (46.1)	
$Re(CO)_6^+$	58.2 (56.0)	52.3 (50.0)	48.4	48.2 (46.0)	50.6 (48.3)	
$Os(CO)_6^{2+}$	69.9 (67.3)	62.6 (60.2)	56.9	58.2 (55.9)	61.0 (58.6)	
$Ir(CO)_6^{3+}$	85.7 (83.4)	79.1 (76.7)	73.7	74.9 (72.6)	77.5 (75.1)	
$Cu(CO)^+$	38.1 (36.7)	51.7 (50.3)		43.3 (41.9)	32.9 (31.5)	36±2
$Cu(CO)^{2+}$	43.1 (41.2)	47.1 (45.2)		42.6 (40.7)	36.7 (34.8)	41±1
$Cu(CO)^{3+}$	23.4 (22.1)	25.0 (23.7)		20.5 (19.2)	19.6 (18.3)	18±1
$Cu(CO)^{4+}$	22.8 (21.0)	21.7 (19.9)		17.3 (15.5)	18.0 (16.2)	13±1
$Ag(CO)^+$	23.3 (22.3)	35.2 (34.2)		29.5 (28.5)	22.0 (21.0)	2±11
$Ag(CO)^{2+}$	28.6 (27.1)	38.4 (36.9)		33.1 (31.6)	26.6 (25.1)	26±1
$Ag(CO)^{3+}$	13.9 (13.3)	16.4 (15.8)		13.8 (13.2)	12.8 (12.2)	13±4
$Ag(CO)^{4+}$	12.3 (11.6)	13.0 (12.3)		11.2 (10.5)	11.3 (10.6)	11(+4/-1)
$Au(CO)^+$	40.8 (39.4)	61.3 (59.9)		49.9 (48.5)	38.5 (37.1)	
$Au(CO)^{2+}$	51.0 (49.0)	57.2 (55.2)		51.9 (49.9)	47.3 (45.3)	
$Au(CO)^{3+}$	9.2 (8.5)	11.5 (10.8)		7.2 (6.5)	6.9 (6.2)	
$Au(CO)^{4+}$	9.3 (8.5)	10.1 (9.3)		7.1 (6.3)	7.3 (6.5)	

[a] BDEs for $M(CO)_6^q$ taken from Refs. 39 and 40.
[b] BDEs for $Cu(CO)_n^+$, $Ag(CO)_n^+$, $Au(CO)_n^+$ taken from Ref. 47.
[c] The calculations of the coinage metal complexes used a relativistic small core ECP with a large valence basis set (311111/22111/411) taken from Ref. 13.
[d] At MP2/II optimized geometries of $M(CO)_6^q$ complexes.
[e] BDEs taken from Ref. 48.

Table 7.2 lists the calculated and experimental BDEs of two series of charged carbonyl complexes [39, 40, 47]. The first series comprises positively and negatively charged hexacarbonyls $M(CO)_6^q$ that are isoelectronic with $W(CO)_6$. There are no experimental bond energies available for $M(CO)_6^q$. The B3LYP/II and BP86/II results are very similar to their CCSD(T)/II counterparts while the MP2/II level of theory always yields values of BDEs that are too high. The second set of data consists of theoretical and experimental results for the positively charged group-11 carbonyls $M(CO)_n^+$ (M = Cu, Ag, Au; n = 1 - 4). The valence basis sets used for the latter metals were much larger than those employed for the hexacarbonyls. Another difference is that the group-11 cations have a completely filled d^{10} shell. The bonding in these compounds has only a negligible M $\rightarrow$ CO π-backdonation, while backdonation is important in the hexacarbonyls. This different bonding situation leads to an altered performance of theoretical methods. As expected, the CCSD(T) values are in good agreement with experiment. The MP2 values are also quite

Table 7.3 Calculated BDEs D_e (D_0) (kcal/mol) of $(CO)_5M{-}L$ complexes.[a]

Molecule	MP2/II	CCSD(T)/II[b]	Exp.
$(CO)_5Cr{-}SiO$	57.7 (56.4)	39.9 (38.6)	
$(CO)_5Mo{-}SiO$	46.4 (45.0)	39.4 (38.0)	
$(CO)_5W{-}SiO$	52.9 (51.3)	45.8 (44.2)	
$(CO)_5Cr{-}CS$	84.9 (82.7)	65.8 (63.6)	58.2±4[c]
$(CO)_5Mo{-}CS$	70.0 (68.4)	60.8 (59.2)	65.6±14[c]
$(CO)_5W{-}CS$	80.3 (78.4)	70.7 (68.8)	71.9±8[c]
$(CO)_5Cr{-}N_2$	33.9 (32.3)	24.8 (23.2)	
$(CO)_5Mo{-}N_2$	26.3 (24.9)	22.0 (20.6)	
$(CO)_5W{-}N_2$	32.6 (31.0)	26.4 (24.8)	
$(CO)_5Cr{-}NO^+$	126.0 (124.7)	106.7 (105.4)	
$(CO)_5Mo{-}NO^+$	123.6 (122.4)	104.4 (103.2)	
$(CO)_5W{-}NO^+$	129.2 (127.8)	110.0 (108.6)	
$(CO)_5Cr{-}CN^-$	108.6 (106.2)	92.1 (89.7)	
$(CO)_5Mo{-}CN^-$	95.4 (93.5)	89.2 (87.3)	
$(CO)_5W{-}CN^-$	106.3 (104.4)	99.6 (97.7)	
$(CO)_5Cr{-}NC^-$	90.8 (89.0)	75.6 (73.8)	
$(CO)_5Mo{-}NC^-$	81.3 (79.8)	76.3 (74.8)	
$(CO)_5W{-}NC^-$	90.8 (89.4)	85.8 (84.4)	
$(CO)_5Cr{-}CCH_2$	93.7 (86.3)[b]	74.1 (66.7)[b]	
$(CO)_5Mo{-}CCH_2$	83.7 (76.0)[b]	70.3 (62.6)[b]	
$(CO)_5W{-}CCH_2$	95.3 (87.6)[b]	80.7 (73.0)[b]	
$(CO)_5Cr{-}H_2$	24.9 (21.0)	19.8 (15.9)	15.0±1.3[d]
$(CO)_5Mo{-}H_2$	17.7 (14.4)	16.1 (12.8)	
$(CO)_5W{-}H_2$	21.9 (18.4)	19.8 (16.3)	$\geq$ 16[d]

[a] BDEs taken from Refs. 42 and 49-51.

[b] At MP2/II optimized geometries.

[c] Ref. 52.

[d] Ref. 53.

accurate. The BP86 and B3LYP functionals yield bond energies for the monocarbonyls $M(CO)^+$ that are much higher than their MP2 and CCSD(T) counterparts. Even more troublesome is the fact that the DFT methods sometimes predict a wrong trend for the BDEs of mono- and dicarbonyls. The BP86 and B3LYP functionals predict the BDE of $Cu(CO)_2^+$ to be lower than that of $Cu(CO)^+$, while MP2 and CCSD(T) levels of theory yield the opposite result, in agreement with experiment.

BP86 also fails to predict the relative BDEs of $Au(CO)^+$ and $Au(CO)_2^+$. The B3LYP approach produces a higher bond energy of $Au(CO)_2^+$ as compared with $Au(CO)^+$ but this difference is much smaller than that predicted within the CCSD(T) approximation.

3. GROUP-6 CARBONYL COMPLEXES $M(CO)_5L$ (M = Cr, Mo, W)

Singly substituted species $M(CO)_{n-1}L$ have been investigated [49-51, 54, 55]. Table 7.3 lists the calculated BDEs of the group-6 complexes of the type $M(CO)_5L$ (M = Cr, Mo, W) with various ligands L, whereas Table 7.4 contains the BDEs for the W–L and the W–CO bonds in $W(CO)_5L$ complexes. The latter values are given for the least bonded carbonyl ligand. Note that BDEs for other $W(CO)_5L$ complexes with certain particular classes of ligands are discussed in other sections of this chapter.

Table 7.3 lists only few experimental data that can be used to estimate the accuracy of the theoretical results. The CCSD(T)/II values agree quite well with experiment (note, however, rather large error bars for the measured BDEs of the thiocarbonyl complexes). The MP2/II values are always larger than their CCSD(T)/II counterparts. The latter values show that the tungsten complexes always have the strongest M–L bond while, in most cases, the molybdenum species have the lowest BDEs.

Table 7.4 lists BDEs calculated with both ab initio and DFT methods. The B3LYP/II values for the W–C_2H_2 and W–$N_2(\sigma)$ bond energies are in very good agreement with their CCSD(T)/II counterparts. However, the results yielded by the two methods differ in the relative bond strengths of acetylene and ethylene in the $W(CO)_5L$ complexes. The B3LYP/II level of theory predicts ethylene to be less bonded than acetylene while CCSD(T)/II and MP2/II levels yield the opposite result. The unstable $W(CO)_5(C_2H_2)$ and $W(CO)_5(C_2H_4)$ species have been detected experimentally by IR spectroscopy [56]. The decrease in the C–O stretching frequency of the trans CO ligand was found to be significantly larger for the former complex. This observation indicates that the tungsten–acetylene interactions are stronger than the tungsten–ethylene ones, which is in agreement with the larger BDE of W–C_2H_2 predicted at the ab initio levels of theory. The only experimental BDE given in Table 7.4 is for the weakly bonded complex $W(CO)_5(CO_2)$. It agrees well with the CCSD(T)/II result.

Table 7.4 Calculated W–L and W–CO BDEs D_e (D_0) (kcal/mol) of $W(CO)_5L$ complexes.[a]

	W–L			W–CO
Molecule	MP2/II	B3LYP/II	CCSD(T)/II[b]	B3LYP/II
$(CO)_5W{-}C_2H_2$ (π)	42.2 (41.2)	33.0 (30.2)	35.3 (34.3)	19.2 (18.1)
$(CO)_5W{-}NCH$ (π)		17.1 (16.2)		30.1 (28.7)
$(CO)_5W{-}NCH$ (σ)		24.7 (23.1)		37.4 (35.0)
$(CO)_5W{-}N_2$ (π)		7.6 (6.7)		41.9 (40.1)
$(CO)_5W{-}N_2$ (σ)	32.6 (31.0)	24.5 (22.7)	26.4 (24.8)	44.9 (42.8)
$(CO)_5W{-}C_2H_4$ (π)	48.3 (46.0)	27.7 (25.4)	41.4 (39.1)	39.8 (37.3)
$(CO)_5W{-}OH_2$ (σ)		25.6 (23.1)		40.2 (38.2)
$(CO)_5W{-}NH_3$ (σ)		35.5 (32.4)		43.7 (41.5)
$(CO)_5W{-}SH_2$ (σ)		22.6 (20.2)		44.1 (42.2)
$(CO)_5W{-}F^-$ (σ)		76.6 (75.9)		25.4 (23.7)
$(CO)_5W{-}Cl^-$ (σ)		55.1 (54.5)		31.5 (29.6)
$(CO)_5W{-}OH^-$ (σ)		82.1 (79.8)		20.2 (19.2)
$(CO)_5W{-}SH^-$ (σ)		63.5 (62.0)		26.9 (25.6)
$(CO)_5W{-}O_2C_2H_2^{2-}$ (σ)		157.1 (154.3)		22.3 (20.6)
$(CO)_5W{-}S_2C_2H_2^{2-}$ (σ)		123.8 (122.6)		24.8 (23.2)
$(CO)_5W{-}CO_2$ (σ)[c]	17.6 (17.1)		10.7 (10.2)	
$(CO)_5W{-}CS_2$ (σ)	31.5 (31.0)		24.6 (24.1)	
$(CO)_5W{-}OCH_2$ (π)	34.1 (32.3)		27.2 (25.4)	

[a] BDEs taken from Refs. 50, 54, and 55.
[b] At MP2/II optimized geometries.
[c] The experimental BDE equals 8.2±1.0 [57].

4. IRON CARBONYL COMPLEXES $Fe(CO)_4L$

A large number of iron carbonyl complexes $Fe(CO)_4L$ with different ligands L in the axial or equatorial positions has been investigated [58]. Table 7.5 lists the theoretically predicted relative energies of the isomers and the Fe–L BDEs at the B3LYP/II and CCSD(T)/II levels of theory. Experimental values of these bond energies are not known. The $Fe(CO)_5L$ complexes, where L is a group-13 diyl ligand, are presented separately.

The data compiled in Table 7.5 show that the relative energies of the axial and equatorial isomers yielded by the B3LYP/II level of theory are very similar to their CCSD(T) counterparts. The B3LYP/II BDEs are always larger than the CCSD(T)/II results, but the trends predicted by the two methods for different ligands are the same. It is worth noting

Table 7.5 Calculated relative energies E_{rel} and and BDEs D_e (D_0) (kcal/mol) with respect to singlet $Fe(CO)_4$ and L (L = CS, N_2, NO^+, CN^-, NC^-, η^2–C_2H_4, η^2–C_2H_2, CCH_2, CH_2^b, CF_2, PH_3, PF_3, NH_3, NF_3, and η^2–H_2).[a]

	B3LYP/II		CCSD(T)/II[d]	
Molecule	E_{rel}[c]	D_e (D_0)	E_{rel}[c]	D_e (D_0)
$(CO)_4Fe$–CS(ax)	0.0	58.4 (55.8)	0.0	66.8 (64.2)
$(CO)_4Fe$–CS(eq)	0.2	58.1 (55.4)	-0.1	66.9 (64.2)
$(CO)_4Fe$–N_2(ax)	0.0	18.7 (16.5)	0.0	25.1 (22.9)
$(CO)_4Fe$–N_2(eq)	1.2	17.5 (15.3)	0.5	24.6 (22.4)
$(CO)_4Fe$–NO^+(ax)	0.0	81.1 (79.2)	0.0	86.7 (84.8)
$(CO)_4Fe$–NO^+(eq)	-13.6	94.7 (92.4)	-20.7	107.4 (105.1)
$(CO)_4Fe$–CN^-(ax)	0.0	89.6 (87.0)	0.0	99.1 (96.5)
$(CO)_4Fe$–CN^-(eq)	6.1	83.5 (81.0)	6.7	92.4 (89.9)
$(CO)_4Fe$–NC^-(ax)	0.0	72.8 (70.7)	0.0	80.7 (78.6)
$(CO)_4Fe$–NC^-(eq)	6.8	66.0 (64.2)	6.3	74.5 (72.7)
$(CO)_4Fe$–(η^2-C_2H_4)(ax)	0.0	21.3 (18.3)	0.0	33.6 (30.6)
$(CO)_4Fe$–(η^2-C_2H_4)(eq)	-7.6	28.9 (25.9)	-8.6	42.2 (39.2)
$(CO)_4Fe$–(η^2-C_2H_2)(ax)	0.0	18.8 (16.9)	0.0	28.8 (26.9)
$(CO)_4Fe$–(η^2-C_2H_2)(eq)	-8.8	27.6 (25.4)	-10.7	39.5 (37.3)
$(CO)_4Fe$–CCH_2(ax)	0.0	68.6 (64.7)	0.0	79.6 (75.7)
$(CO)_4Fe$–CCH_2(eq)	-6.2	74.8 (70.3)	-8.7	88.3 (83.8)
$(CO)_4Fe$–CH_2(ax)	0.0	74.3 (69.1)	0.0	84.8 (79.6)
$(CO)_4Fe$–CH_2(eq)	-6.5	80.8 (75.1)	-8.3	93.1 (87.4)
$(CO)_4Fe$–CF_2(ax)	0.0	55.1 (52.2)	0.0	62.7 (59.8)
$(CO)_4Fe$–CF_2(eq)	-3.0	58.2 (55.2)	-4.6	67.3 (64.3)
$(CO)_4Fe$–NH_3(ax)	0.0	33.7 (29.9)	0.0	42.9 (39.1)
$(CO)_4Fe$–NH_3(eq)	6.6	27.1 (23.7)	6.4	36.5 (33.1)
$(CO)_4Fe$–NF_3(ax)	0.0	16.9 (15.3)	0.0	25.1 (23.5)
$(CO)_4Fe$–NF_3(eq)	3.4	13.5 (12.2)	3.0	22.2 (20.9)
$(CO)_4Fe$–PH_3(ax)	0.0	30.2 (26.8)	0.0	42.3 (38.9)
$(CO)_4Fe$–PH_3(eq)	1.1	29.1 (25,9)	2.7	39.7 (36.5)
$(CO)_4Fe$–PF_3(ax)	0.0	36.6 (34.1)	0.0	47.6 (45.2)
$(CO)_4Fe$–PF_3(eq)	-0.3	36.9 (34.5)	1.0	46.5 (44.1)
$(CO)_4Fe$–(η^2-H_2)(ax)	0.0	15.0 (10.3)	0.0	21.2 (16.5)
$(CO)_4Fe$–(η^2-H_2)(eq)	-2.0	17.1 (12.8)	-1.6	22.8 (18.5)

[a] Relative energies and BDEs taken from Ref. 58.
[b] Triplet CH_2.
[c] Relative to the axial isomer and without ZPE correction.
[d] At B3LYP/II optimized geometries.

that the B3LYP/II and CCSD(T)/II levels of theory agree on ethylene in $Fe(CO)_4L$ being more strongly bonded than acetylene.

Table 7.6 Calculated BDEs D_e (D_0) (kcal/mol) of $(CO)_3M–L$ complexes.[a]

Molecule	MP2/II	CCSD(T)/II
$(CO)_3Ni–SiO$	41.8 (40.8)	21.3 (20.3)
$(CO)_3Pd–SiO$	23.4 (23.3)	16.0 (15.9)
$(CO)_3Pt–SiO$	33.0 (32.0)	23.2 (22.2)
$(CO)_3Ni–CS$	70.2 (69.1)	36.8 (35.7)
$(CO)_3Pd–CS$	29.7 (29.1)	18.6 (18.0)
$(CO)_3Pt–CS$	41.9 (40.8)	28.0 (26.9)
$(CO)_3Ni–N_2$[b]	27.5 (26.8)	5.3 (4.6)
$(CO)_3Ni–NO^+$	108.6 (107.7)	69.8 (68.9)
$(CO)_3Pd–NO^+$	69.2 (68.8)	51.4 (51.0)
$(CO)_3Pt–NO^+$	66.0 (65.1)	46.6 (45.7)
$(CO)_3Ni–CN^-$	89.3 (88.0)	61.7 (60.4)
$(CO)_3Pd–CN^-$	54.6 (53.8)	45.7 (44.9)
$(CO)_3Pt–CN^-$	96.0 (94.5)	58.6 (57.1)
$(CO)_3Ni–NC^-$	72.9 (72.3)	47.7 (47.1)
$(CO)_3Pd–NC^-$	40.0 (39.9)	33.0 (32.9)
$(CO)_3Pt–NC^-$	47.9 (47.3)	39.3 (38.7)
$(CO)_3Ni–CCH_2$		(38.8)
$(CO)_3Pd–CCH_2$		(22.2)
$(CO)_3Pt–CCH_2$		(33.5)

[a] BDEs taken from Refs. 19 and 49.

[b] The estimated experimental value is 10.7 kcal/mol [59].

5. GROUP-10 CARBONYL COMPLEXES $M(CO)_3L$ (M = Ni, Pd, Pt)

Table 7.6 lists the theoretical BDEs of the M–L bonds in the group-10 $Ni(CO)_3L$, $Pd(CO)_3L$ and $Pt(CO)_3L$ complexes calculated at the MP2/II and CCSD(T)/II levels of theory [49, 50]. The only experimental value known for those compounds is an estimate of ca. 10 kcal/mol obtained for the $(CO)_3Ni–N_2$ bond energy at 298 K [59]. This estimate is based on kinetic measurements of nitrogen extrusion from the complex. Thermal corrections to the CCSD(T)/II value of D_0 = 4.6 kcal/mol yield a theoretical prediction of 6.7 kcal/mol, which is in a reasonable agreement with experiment [49]. The MP2/II BDEs listed in

Table 7.6 are significantly larger than their CCSD(T)/II counterparts. $Ni(CO)_3N_2$ is much more strongly bonded at the MP2/II level of theory (D_0 = 26.8 kcal/mol) than at the CCSD(T)/II level. The stronger metal–ligand bonds at the MP2/II level of theory lead to energy minima for the $M(CO)_3L$ complexes that are predicted by the CCSD(T)/II calculations to be unstable with respect to the M–L dissociation. For example, both $Pd(CO)_3(C_2H_2)$ and $Pt(CO)_3(C_2H_2)$ are minima on the MP2/II potential energy hypersurfaces, while the calculations at the CCSD(T)/II level of theory produce negative dissociation energies [42]. The M–L BDEs of the group-10 carbonyl complexes $M(CO)_3L$ are clearly lower than those of their group-6 analogues $M(CO)_5L$. They are also lower than the Fe–L BDEs of $Fe(CO)_4L$.

6. GROUP-6 CARBONYL COMPLEXES WITH PHOSPHANE LIGANDS $M(CO)_5PR_3$ (M = Cr, Mo, W; R = H, Me, F, Cl)

The structure and bonding of group-6 TM carbonyl complexes $M(CO)_5PR_3$ with phosphane ligands PH_3, PMe_3, PF_3, and PCl_3 have been the subjects of another theoretical study [60]. Table 7.7 lists the M–PR_3 BDEs calculated at the BP86 level of theory in conjunction with our standard basis set II and the larger TZ(2)P Slater basis set, which has one set of *f*-type polarization functions on the transition metals and two sets of polarization functions on the other atoms.

Except for the PF_3 complexes, the BP86/II and BP86/TZ(2)P BDEs values are very similar. The BP86/II BDE estimates for these complexes are 4 - 6 kcal/mol higher than their BP86/TZ(2)P counterparts. Both levels of theory predict the trend in the M–PR_3 BDEs for the different phosphane ligands being $PCl_3 < PH_3 < PF_3 < PMe_3$.

7. NOBLE GAS COMPLEXES $M(CO)_5Ng$ (M = Cr, Mo, W; Ng = Ar, Kr, Xe)

A special type of TM ligands are the noble gas atoms argon, krypton, and xenon [61]. Although they are weak Lewis bases, TM complexes $M(CO)_5Ng$ with M = Cr, Mo, W and Ng = Ar, Kr and Xe have been experimentally investigated in the gas phase as well as in the liquid phase and in supercritical CO_2 [62, 63]. The M–Ng BDEs were estimated with

Table 7.7 Calculated BDEs D_e (D_0) (kcal/mol) of phosphane complexes $M(CO)_5PR_3$ (M = Cr, Mo, W; R = H, Me, F, Cl).[a]

Molecule	BP86/II	BP86/TZ(2)P
$(CO)_5Cr–PH_3$	31.7 (29.0)	31.9 (29.2)
$(CO)_5Mo–PH_3$	29.8 (27.5)	29.6 (27.3)
$(CO)_5W–PH_3$	35.3 (33.0)	37.2 (34.8)
$(CO)_5Cr–PMe_3$	41.8 (39.8)	41.4 (39.4)
$(CO)_5Mo–PMe_3$	39.0 (37.6)	37.8 (36.5)
$(CO)_5W–PMe_3$	45.5 (44.1)	43.8 (42.4)
$(CO)_5Cr–PF_3$	38.8 (37.0)	33.7 (31.9)
$(CO)_5Mo–PF_3$	37.2 (35.9)	31.4 (30.0)
$(CO)_5W–PF_3$	43.3 (41.9)	39.6 (38.2)
$(CO)_5Cr–PCl_3$	28.5 (29.0)	26.5 (25.1)
$(CO)_5Mo–PCl_3$	27.2 (26.2)	24.6 (23.6)
$(CO)_5W–PCl_3$	32.6 (31.6)	33.3 (32.2)

[a] BDEs taken from Ref. 60.

different techniques. Thus it has become possible to investigate the performance of various theoretical methods in calculations of BDEs for weakly bonded ligands. The experimental and calculated BDEs are compiled in Table 7.8. The calculated values at both the BP86/TZP and CCSD(T)/II levels of theory agree reasonably well with the experimental results. The basis set superposition error (BSSE) corrections to the CCSD(T) values are much smaller than those to BP86. Theory and experiment agree that the BDEs exhibit the trend Ar < Kr < Xe that should be expected on the ground of the electric polarizabilities of these elements.

8. TRANSITION METAL CARBENE AND CARBYNE COMPLEXES

TM complexes with carbene ligands $:CR_2$ are particularly interesting from a theoretical standpoint. Two classes of such species, for which different bonding models have been suggested, namely the Fischer and Schrock complexes, are known [3, 64]. The bonding situation in the members of the former class can be described in terms of donor-acceptor interactions between the metal fragment and a singlet carbene analogous

Table 7.8 Calculated BDEs (kcal/mol) of noble gas complexes $(CO)_5M$–Ng (Ng = Ar, Kr, Xe).[a]

	BP86/TZP		CCSD(T)/II		
Molecule	D_e (D_0^{298})	$D_0^{298\,d}$	D_e (D_0^{298})	$D_0^{298\,d}$	Exp.
$(CO)_5Cr$–Ar	4.9 (6.3)	3.5	1.9 (3.3)	3.0	
$(CO)_5Mo$–Ar	5.4 (6.8)	2.2	2.7 (4.1)	3.6	
$(CO)_5W$–Ar	8.0 (9.4)	4.3	3.6 (5.0)	4.6	ca. $\leq 3^b$
$(CO)_5Cr$–Kr	6.2 (7.5)	4.7	3.0 (4.3)	4.0	
$(CO)_5Mo$–Kr	6.9 (8.2)	4.4	3.9 (5.2)	4.7	
$(CO)_5W$–Kr	10.0 (11.3)	6.7	5.1 (6.4)	6.0	$< 6^b$
$(CO)_5Cr$–Xe	7.2 (8.5)	5.0	5.4 (6.7)	6.4	9.0 ± 0.9^b
$(CO)_5Mo$–Xe	8.2 (9.5)	4.7	7.0 (8.3)	7.9	8.0 ± 1.0^b
$(CO)_5W$–Xe	11.9 (13.2)	7.6	7.6 (8.9)	8.8	8.2 ± 1.0^b
					8.4 ± 0.2^c

[a] BDEs taken from Ref. 61.

[b] Ref. 62.

[c] Ref. 63.

[d] Corrected for BSSE.

to those involved by the familiar Dewar-Chatt-Duncanson (DCD) model of olefin complexes. Metal–carbene bonding in Schrock complexes is better understood by electron-sharing covalent bonding between unpaired electrons of the metal and a triplet carbene. Quantum-chemical studies of the bonding interactions in TM carbene complexes have been reported [3, 65].

Table 7.9 lists the calculated BDEs of tungsten carbene complexes that belong to either class of compounds [65, 66]. The $W(CO)_5CR_2$ complexes are of the Fischer type while the WX_4CR_2 species are of the Schrock type. Analysis of the bonding interactions in the $WF_5(CR_2)$ molecules indicates that they possess W–CR_2 donor-acceptor bonds, i.e. they are Fischer complexes [65]. The complexes of MCl (M = Cu, Ag, Au) with Arduengo-type carbenes and their heavier homologues exhibit yet another type of bonding. These species have mainly metal←ligand π donation and negligible metal→ligand π backdonation [66]. This distinguishes them from Fischer carbene complexes, where the metal→carbene π backdonation constitutes an important part of the bonding.

Table 7.9 Calculated BDEs D_e (D_0) (kcal/mol) of complexes with carbene ligands and their heavier homologues.[a]

Metal fragment	Ligand	MP2/II	CCSD(T)/II[b]
$W(CO)_5$	CH_2	81.3 (75.7)	78.9 (73.3)
$W(CO)_5$	CF_2	67.5 (65.7)	60.6 (58.8)[c]
$W(CO)_5$	CHF	86.9 (84.1)	80.0 (77.2)[c]
$W(CO)_5$	CH(OH)	81.9 (78.0)	75.0 (71.1)[c]
WF_4	CH_2	127.7 (125.6)	118.2 (116.1)
WCl_4	CH_2	91.2 (89.5)	75.3 (73.6)
WBr_4	CH_2	87.9 (86.4)	74.2 (72.7)
WI_4	CH_2	82.5 (81.4)	70.5 (69.4)
WF_4	CF_2	65.3 (63.7)	57.5 (55.9)
WF_5^-	CH_2	110.0 (104.5)	101.3 (95.8)
WF_5^-	CF_2	70.7 (68.6)	62.8 (60.7)
CuCl	$C_3H_4N_2$	76.3	67.4
AgCl	$C_3H_4N_2$	61.2	56.5
AuCl	$C_3H_4N_2$	88.6	82.8
CuCl	$SiC_2H_4N_2$	52.7	45.1
AgCl	$SiC_2H_4N_2$	41.5	37.4
AuCl	$SiC_2H_4N_2$	68.5	64.1
CuCl	$GeC_2H_4N_2$	39.5	35.1
AgCl	$GeC_2H_4N_2$	33.0	29.9
AuCl	$GeC_2H_4N_2$	53.5	49.4

[a] BDEs for the tungsten complexes taken from Ref. 65;
BDEs for the group-11 complexes taken from Ref. 66.
[b] At MP2/II optimized geometries.
[c] BDEs estimated from isostructural reactions [19].

Inspection of Table 7.9 leads to the conclusion that the bonds in these three classes of compounds are very strong. The bond energy can be significantly altered by the nature of the substituent R in the $:CR_2$ ligand. This is an important piece of information as there are no experimental data available for the BDEs of stable carbene complexes. The MP2/II BDEs are always higher than their CCSD(T)/II counterparts, although not by much. The trends predicted by the two methods for different ligands and different coinage metals are essentially the same.

The same dichotomy of bonding models is also found for carbyne complexes that have a formal triple bond $M{\equiv}CR$. There are metal–carbyne bonds that belong to the donor-acceptor type (the Fischer car-

Table 7.10 Calculated BDEs D_e (D_0) (kcal/mol) of transition metal carbyne complexes.[a]

Metal fragment	Ligand	MP2/II	CCSD(T)/II
$W(CO)_4Br$	CH	155.0 (149.2)	133.9 (128.1)
	CF	122.5 (118.9)	105.0 (101.4)
	$C(NH_2)$	110.9 (107.4)	95.2 (91.7)
WCl_3	CH	177.5 (172.0)	154.5 (149.0)
	CF	133.0 (127.5)	111.7 (106.2)
	$C(NH_2)$	124.0 (120.0)	104.9 (100.9)
WCl_4^-	CH	205.9 (202.8)	176.5 (173.4)

[a] BDEs taken from Ref. 67.

bynes) the others can be more conveniently analyzed in terms of electron-sharing triple bonds between open-shell fragments (the Schrock carbynes) [3]. The M–CR BDEs of the Fischer-type carbyne complexes $Br(CO)_4W{-}CR$ and the Schrock-type species $Cl_3W{\equiv}CR$ and $Cl_4W{\equiv}CR^-$ have been calculated [67], and are compiled in Table 7.10. These bond energies are much higher than those computed for analogous carbene complexes. It is also worth noting that the M–CR BDEs become significantly smaller for R being good π-electron donors.

9. TRANSITION METAL COMPLEXES WITH π-BONDED LIGANDS

The dichotomy of donor-acceptor versus electron-sharing bonding models also gives rise to an important classification scheme for TM complexes with π bonded ligands. The bonding between a TM and an olefin can be understood either in terms of the DCD model or as bonding in a metallacyclopropane. Compounds with alkyne ligands can likewise be described as alkyne complexes or metallacyclopropenes. Therefore, in general, ligands with high-lying occupied π orbitals can be divided into donor-acceptor complexes and metallacyclic compounds, respectively.

Table 7.11 lists the predicted BDEs of TM compounds with π-donor ligands [4, 54, 55, 68-71]. The complexes of $W(CO)_5$ with acetylene, ethylene, and formaldehyde belong to the donor-acceptor class. The compounds of WCl_4 with the same ligands are metallacyclic molecules.

Table 7.11 Calculated BDEs D_e (D_0) (kcal/mol) of transition metal complexes with π-bonded ligands.[a]

Metal fragment	π-Ligand	B3LYP/II	MP2/II	CCSD(T)/II
$W(CO)_5$	C_2H_2	33.0 (30.2)	42.2 (41.2)	35.3 (34.3)
	C_2H_4	27.7 (15.4)	48.3 (46.0)	41.4 (39.1)
	CH_2O		34.1 (32.3)	27.2 (25.4)
WCl_4	C_2H_2		52.1 (49.9)	36.6 (34.4)
	C_2H_4		22.4 (19.9)	12.1 (9.6)
	CS_2		13.9 (13.0)	-3.2 (-4.1)
	CH_2O		35.6 (32.9)	18.7 (16.0)
WCl_5^-	C_2H_2		38.3 (36.2)	22.3 (20.2)
	C_2H_4		17.8 (15.0)	8.5 (5.7)
$Pt(PH_3)_2$	C_2H_4			23.9 (22.3)
	$C_{11}H_{16}$[b]			35.2 (34.7)
	$C_{10}H_{14}$[b]			47.9 (46.9)
	C_9H_{12}[b]			58.5 (57.4)
	C_8H_{10}[b]			70.1 (69.1)
$CuCH_3$	C_2H_2	20.0 (22.4)		
	$(C_2H_2)_2$	17.4 (18.3)		
	$H_2Si(CCH)_2$	15.3 (15.9)		
	$H_2C(CCH)_2$	11.5 (12.1)		
Cu^+	C_2H_2		45.5 (44.1)[c]	40.6 (39.2)[d]
	C_2H_4		48.8 (46.8)[c]	43.9 (41.9)[d]
Cu	C_2H_2		7.5 (4.4)[c]	2.3 (-0.8)[d]
	C_2H_4		7.9 (0.0)[c]	4.2 (-3.7)[d]

[a] BDEs taken from Refs. 54, 55, and 68-71.

[b] A strained tricyclic alkene [69].

[c] Computed using relativistic ECP with a (311111/2211/411) valence basis set for Cu and 6-31G(d) for C and H.

[d] Computed using a relativistic ECP with a (311111/2211/411/1) valence basis set for Cu and 6-31G+G(d) for C and H.

The CCSD(T)/II calculations provide similar values for the $(CO)_5W$–C_2H_2 and Cl_4W–C_2H_2 BDEs. The $(CO)_5W$–C_2H_4 species is predicted to have higher BDE than $(CO)_5W$–C_2H_2, whereas Cl_4W–C_2H_4 is clearly less bonded than Cl_4W–C_2H_2. This change in trend can be explained by the different bonding situations in metallacyclic compounds and donor-acceptor complexes [4]. At the MP2/II level of theory, CS_2 in Cl_4W–

CS_2 is found to bind to tungsten in a π-type fashion. However, the CCSD(T)/II calculations predict the compound to be thermodynamically unstable relative to its components in their electronic ground states. The $W(CO)_5CO_2$ and $W(CO)_5CS_2$ complexes possess σ-bonded ligands (see Table 7.4), whereas $Cl_5W{-}C_2H_2^-$ and $Cl_5W{-}C_2H_4^-$ are π-bonded species.

The BDEs of the $(PH_3)_2Pt{-}L$ platinum complexes, where the ligand L is an olefin, have also been calculated [69]. Table 7.11 lists these BDEs computed at the CCSD(T)/II level of theory. The least bonded ligand is ethylene. The other ligands are strained tricyclic olefines $R_2C{=}CR_2$ with pyramidal carbon skeletons that raise their HOMOs and lower their LUMOs. The strain angle increases from $C_{11}H_{16}$ to C_8H_{10}, enhancing the metal–olefin interactions in this direction. Table 7.11 shows that the BDE increases when the olefin becomes more strongly pyramidalized and therefore due to the ring strain more prone to bond formation.

Table 7.11 also contains the BDEs of naked Cu and Cu^+ with acetylene and ethylene as ligands [68]. The bond energy of the positively charged copper cation is, as expected, much higher than that of the neutral atom. It is worth pointing out here that the higher BDE of $Cu(C_2H_n)^+$ is not reflected by the calculated C–C bond lengths, which are nearly the same in both cases [68]. This can be explained on the grounds of the metal–ligand bonding situation that in the $Cu(C_2H_n)^+$ cations is mainly ionic. The bond path provided by the topological analysis of the electron density is T-shaped rather than cyclic [68].

10. TRANSITION METAL COMPLEXES WITH GROUP-13 DIYL LIGANDS ER (E = B, Al, Ga, In, Tl)

Apart from carbonyl complexes, another very extensively investigated class of TM compounds is that of complexes attached to group-13 diyl ligands ER (E = B - Tl) with various substituents R [6, 72-79]. The aim of these investigations was to gain understanding of the metal–ligand interactions in the compounds that have only recently been synthesized and characterized by X-ray structure analysis [72]. As a side-product of the bonding analysis a large number of M–ER BDEs was calculated. These BDEs are presented in Tables 7.12 and 7.13.

Table 7.12 lists the theoretical BDEs of the $(CO)_4Fe{-}ER$ bonds calculated at the BP86/II and BP86/TZP levels of theory. The calculations predict the following trend of the bond strengths of the ligands: EMe $\approx$ EPh > $EN(SiH_3)_2$ > ECp. The BP86/TZP values suggest for

Table 7.12 Calculated BDEs D_e (D_0) (kcal/mol) of iron carbonyl complexes with group-13 diyl ligands ER (E = B - Tl).[a]

Metal fragment	Ligand	BP86/II	BP86/TZP
$Fe(CO)_4$	BCp (ax)	78.0 (75.0)	75.3 (72.3)
	BCp (eq)	[b]	70.0
	AlCp (ax)	53.1 (51.4)	52.7 (51.0)
	AlCp (eq)	52.4 (50.5)	51.7 (49.8)
	GaCp (ax)	32.9 (31.6)	23.0 (21.7)
	GaCp (eq)	32.4 (31.0)	22.9 (21.5)
	InCp (ax)	33.9 (31.7)	19.8 (17.6)
	InCp (eq)	33.3 (32.0)	20.0 (18.7)
	TlCp (ax)	16.7 (15.8)	13.6 (12.7)
	TlCp (eq)	17.1 (16.1)	13.1 (12.1)
$Fe(CO)_4$	$BN(SiH_3)_2$ (ax)	[b]	83.6
	$BN(SiH_3)_2$ (eq)	85.8 (83.1)	83.4 (80.7)
	$AlN(SiH_3)_2$ (ax)	51.8 (50.2)	51.6 (50.0)
	$AlN(SiH_3)_2$ (eq)	53.0 (51.2)	51.5 (49.7)
	$Ga(SiH_3)_2$ (ax)	39.7 (38.2)	34.5 (33.0)
	$Ga(SiH_3)_2$ (eq)	39.5 (37.9)	33.6 (32.0)
	$InN(SiH_3)_2$ (ax)	38.9 (37.5)	28.9 (27.5)
	$InN(SiH_3)_2$ (eq)	38.2 (36.7)	27.5 (26.0)
	$TlN(SiH_3)_2$ (ax)	25.4 (24.2)	20.7 (19.5)
	TlN(SiH3)2 (eq)	25.0 (22.5)	20.4 (17.9)
$Fe(CO)_4$	BPh (ax)	102.8 (99.8)	100.2 (97.2)
	BPh (eq)	[b]	99.0
	AlPh (ax)	63.5 (61.5)	63.8 (61.8)
	AlPh (eq)	63.6 (61.5)	62.9 (60.8)
	GaPh (ax)	55.0 (53.2)	52.3 (50.5)
	GaPh (eq)	53.3 (51.4)	49.4 (47.5)
	InPh (ax)	53.2 (51.5)	40.7 (39.0)
	InPh (eq)	51.3 (49.4)	43.5 (41.6)
	TlPh (ax)	42.5 (41.0)	40.8 (39.3)
	TlPh (eq)	40.0 (38.4)	38.0 (36.4)
$Fe(CO)_4$	BMe (ax)		100.0
	BMe (eq)		98.1
	AlMe (ax)		65.4
	AlMe (eq)		64.6
	GaMe (ax)		53.8
	GaMe (eq)		50.8
	InMe (ax)		48.4
	InMe (eq)		45.5
	TlMe (ax)		45.8
	TlMe (eq)		46.4
	InCl	33.2	
	$InCl_2^-$	64.7	

[a] BDEs taken from Refs. 76 and 79.

[b] Not an energy minimum at this level of theory.

Table 7.13 Calculated BDEs D_e (D_0) (kcal/mol) of homoleptic iron and group-10 transition metal diyl complexes.[a]

Metal fragment	Ligand	B3LYP/II	BP86/II	BP86/TZP
$Fe(BCH_3)_4$	BCH_3			105.6
$Fe(AlCH_3)_4$	$AlCH_3$			79.2
$Fe(GaCH_3)_4$	$GaCH_3$			64.1
$Fe(InCH_3)_4$	$InCH_3$			57.4
$Fe(TlCH_3)_4$	$TlCH_3$			53.1
$Ni(BCH_3)_3$	BCH_3	83.8 (79.7)	91.4 (87.3)	92.3 (88.2)
$Ni(AlCH_3)_3$	$AlCH_3$	55.6 (53.2)	61.6 (59.2)	61.6 (59.2)
$Ni(GaCH_3)_3$	$GaCH_3$	43.2 (40.7)	49.6 (47.1)	46.6 (44.1)
$Ni(InCH_3)_3$	$InCH_3$	45.4 (43.4)	51.1 (49.1)	40.7 (38.7)
$Ni(TlCH_3)_3$	$TlCH_3$	28.4 (27.0)	[b]	35.7
$Pd(BCH_3)_3$	BCH_3	67.5 (63.8)	76.2 (72.5)	
$Pd(AlCH_3)_3$	$AlCH_3$	46.0 (43.6)	52.9 (50.5)	
$Pd(GaCH_3)_3$	$GaCH_3$	33.4 (31.3)	40.3 (38.2)	
$Pd(InCH_3)_3$	$InCH_3$	37.4 (35.4)	43.7 (41.7)	
$Pd(TlCH_3)_3$	$TlCH_3$	19.9 (18.5)	26.8 (25.4)	
$Pt(BCH_3)_3$	BCH_3	82.7 (79.0)	89.3 (85.6)	
$Pt(AlCH_3)_3$	$AlCH_3$	57.3 (54.8)	62.8 (60.3)	
$Pt(GaCH_3)_3$	$GaCH_3$	43.3 (41.2)	48.9 (46.7)	
$Pt(InCH_3)_3$	$InCH_3$	46.8 (44.8)	51.9 (49.9)	
$Pt(TlCH_3)_3$	$TlCH_3$	26.2 (24.7)	32.1 (28.6)	
Pt(dhpe)(AlCp)[c]	AlCp		29.6	
Pt(dhpe)(GaCp)[c]	GaCp		18.3	

[a] BDEs taken from Refs. 76, 78, and 79.

[b] No SCF convergence.

[c] dphe = diphosphinoethane. A larger valence set (211111/411/41) was used for Pt in conjunction with the SVP basis set for the other atoms.

the different ligand atoms E the following trend in the BDEs: BR > AlR > GaR > InR > Tl. The BP86/II data predict the same trend except that the calculated BDEs of the GaR and InR ligands are nearly the same. It is worth noting that the BP86/II and BP86/TZP results for boron and aluminum complexes are very similar, while for the gallium, indium and thallium complexes, the BP86/II level of theory predicts much higher bond energies than does BP86/TZP. This is probably due to the difference in the II basis sets that are used for B and Al and those that are used for Ga, In and Tl. The former group of elements is treated

Table 7.14 Calculated BDEs D_e (D_0) (kcal/mol) of $(CO)_5W$–L complexes with group-13 diyl ligands L.[a]

Metal fragment	Ligand	MP2/II	BP86/II
$W(CO)_5$	AlH	70.0 (68.1)	
$W(CO)_5$	$AlH(NH_3)_2$	100.9 (97.9)	
$W(CO)_5$	AlCl	58.4 (57.4)	40.5 (39.5)
$W(CO)_5$	$AlCl(NH_3)_2$	93.1 (86.2)	69.2 (62.3)
$W(CO)_5$	BCl		73.8
$W(CO)_5$	$BCl(NH_3)_2$	119.6	88.8
$W(CO)_5$	$GaCl(NH_3)_2$	70.9	48.5
$W(CO)_5$	$InCl(NH_3)_2$	70.5	49.7
$W(CO)_5$	$TlCl(NH_3)_2$	47.8	28.6

[a] BDEs taken from Refs. 73 and 80.

with the all-electron 6-31G(d) basis sets, whereas ECPs associated with (31/31/1) valence basis sets are employed for the heavier group-13 atoms. Because of the above procedure, we believe that, being obtained with the same type of basis sets for all atoms, the BP86/TZP values are more reliable.

Inspection of Table 7.13 leads to the conclusion that the Fe–EMe BDEs of the homoleptic complexes $Fe(EMe)_5$ are higher than their counterparts predicted for the $(CO)_4Fe$–EMe bonds. The homoleptic group-13 complexes $M(EMe)_4$ are also predicted to have high bond energies, except for the Pt–AlCp and Pt–GaCp BDEs in $Pt(dhpe)(ECp)_2$ that are rather low.

A number of tungsten group-13 diyl complexes $(CO)_5W$–ER (R = H, Cl) and $(CO)_5W$–$ER(NH_3)_2$ have also been investigated. Thanks to the stabilizing effect of the ammonia ligands, the latter complexes can be isolated and their structures can be determined with X-ray analysis [73]. The ammonia-free $(CO)_5W$–ER species has not been isolated yet. It is worth emphasizing that, as can be seen from Table 7.14, the addition of ammonia to the diyl ligands strongly enhances the W–E bond energy.

Table 7.15 Calculated BDEs D_e (D_0) (kcal/mol) of transition metal complexes with boryl ligands BR_2[a] and gallyl ligands $GaR_2(NH_3)$[b].

Molecule	B3LYP/II	
	TM–ER_2	TM–CO[c]
$Os(PH_3)_2(CO)Cl(BH_2)$	90.0 (85.6)	
$Os(PH_3)_2(CO)Cl(BF_2)$	90.8 (88.3)	
$Os(PH_3)_2(CO)Cl(B(OH)_2)$	84.7 (81.9)	
$Os(PH_3)_2(CO)Cl(B(OHC{=}CHO))$	88.1 (86.1)	
$Os(PH_3)_2(CO)Cl(Bcat)$[d]	87.1 (85.3)	
$Os(PH_3)_2(CO)_2Cl(BH_2)$	78.1 (72.7)	26.8 (22.2)
$Os(PH_3)_2(CO)_2Cl(BF_2)$	80.5 (77.9)	27.0 (24.7)
$Os(PH_3)_2(CO)_2Cl(B(OH)_2)$	77.4 (74.4)	30.2 (27.7)
$Os(PH_3)_2(CO)_2Cl(B(OHC{=}CHO))$	79.3 (77.1)	28.5 (26.1)
$Os(PH_3)_2(CO)_2Cl(Bcat)$[d]	77.8 (75.8)	28.1 (25.7)
$(CO)_4CoGaH_2(NH_3)$	123.1	
$(CO)_4CoGaCl_2(NH_3)$	143.3	
$(PH_3)(CO)_3CoGaCl_2(NH_3)$	157.5	

[a] BDEs taken from Ref. 82.
[b] BDEs taken from Ref. 81.
[c] CO ligands trans to the BR_2 group.
[d] cat = catecholate $O_2C_6H_4^{2-}$.

11. TRANSITION METAL COMPOUNDS WITH BORYL LIGANDS BR_2 AND GALLYL LIGANDS GaR_2

In addition to the complexes with group-13 diyl ligands ER, where the element E is in the formal oxidation state I, TM compounds with electron-sharing M–ER_2 bonds, where the element E is in the formal oxidation state III, have been investigated [81, 82]. Table 7.15 lists the calculated Os–BR_2 BDEs of osmium complexes that are 16-valence electron species with the general formula $Os(PH_3)_2(CO)Cl(BR_2)$. The BDEs of the Os–BR_2 and Os–CO_{trans} bonds of the 18-valence electron $Os(PH_3)_2(CO)_2Cl(BR_2)$ complexes have been investigated as well. The above compounds are regarded as models for the stable 16- and 18-valence electron species with bulky PR_3 ligands [83].

The calculations indicate that the Os–BR_2 BDEs of the latter complexes are 10 - 12 kcal/mol lower than those of the former ones. The second carbonyl ligand that is in the trans position with respect to the boryl group in $Os(PH_3)_2(CO)_2Cl(BR_2)$ has a rather low BDE. This result may be expected since both classes of complexes coexist and can be interconverted by changing the CO pressure [83]. Table 7.15 also lists the calculated Co–$GaR_2(NH_3)$ BDEs. These bond energies are significantly higher that those of the osmium compounds, although boron generally forms stronger bonds with transition metals than gallium. The high bond energy may partly be explained by the effect of the ammonia ligand that has been already shown to enhance the bond strength of the M–E bond, when M is a group-13 element.

12. TRANSITION METAL METHYL AND PHENYL COMPOUNDS

Another class of M–ligand bonds for which BDEs have been calculated are methyl and phenyl bonds of the group-11 metals Cu, Ag, Au and the group-12 metals Zn, Cd and Hg [28]. Table 7.16 compiles the predicted BDEs calculated at the MP2/II and CCSD(T)/II levels of theory as well as the relevant experimental data.

The most interesting conclusion from the results shown in Table 7.16 is that, except for $Hg(C_6H_5)_2$, the MP2/II BDEs are in excellent agreement with the experimental data. Although the good agreement between the MP2/II calculations and experiment is due to a fortuitous error cancellation, the question remains why the theoretical value for $Hg(C_6H_5)_2$ is too high when all the other values are correct. The calculations predict that the metal–phenyl bonds have higher BDEs than the metal–methyl bonds, while the only experimental value available for a phenyl compound of a group-12 element predicts the opposite. Since sp^2-hybridized bonds of carbon are normally stronger than their sp^3-hybridized counterparts, we tend to believe that the theoretical value is correct. Unfortunately, CCSD(T)/II calculations of $Hg(C_6H_5)_2$ could not be carried out because the molecule was too big. The BDEs predicted at the CCSD(T)/II level of theory are slightly lower than both their MP2/II and experimental counterparts but the trend is the same.

Table 7.16 Calculated BDEs D_e (D_0) (kcal/mol) of transition metal–methyl and metal–phenyl complexes.[a]

Molecule	MP2[b]	CCSD(T)[b]	Exp.
$CuCH_3$	58.7 (55.5)	52.1 (48.9)	53.3[c]
$AgCH_3$	43.1 (39.9)	40.8 (37.6)	
$AuCH_3$	62.5 (58.6)	58.1 (54.0)	
CuC_6H_5[b]	74.0 (72.2)	65.1 (63.3)	
AgC_6H_5[b]	58.3 (56.7)	53.8 (52.2)	
AuC_6H_5[b]	80.2 (78.2)	72.4 (70.4)	
$Zn(CH_3)_2$	90.4 (83.1)	81.1 (73.8)	84.8[e]
$Cd(CH_3)_2$	73.8 (66.7)	67.8 (60.7)	67.2[e]
$Hg(CH_3)_2$	66.6 (58.7)	60.5 (52.6)	57.9[e]
$Zn(C_6H_5)_2$[d]	123.2 (118.9)		
$Cd(C_6H_5)_2$[d]	109.2 (105.2)		
$Hg(C_6H_5)_2$[d]	103.3 (99.0)		75.3[e]

[a] BDEs taken from Ref. 28.

[b] For details on the basis sets employed see Ref. 28.

[c] Ref. 84.

[e] BDEs derived from the respective differences in the values of ΔH_f^0 [85]; see Ref. 28 for details.

13. TRANSITION METAL NITRIDO AND PHOSPHIDO COMPLEXES

The nitrido and phosphido complexes of TMs have been the subjects of intensive experimental studies in the recent years. Of particular interest has been the issue of Lewis basicity of the nitrogen and phosphorus atoms in the TM≡N and TM≡P groups. Table 7.17 lists the BDEs calculated at the MP2/II, B3LYP/II and CCSD(T)/II levels of theory for L_nMN–X and L_nMP–X, where X is a group-13 Lewis acid or a chalcogen atom [86, 87].

These results demonstrate that it is difficult to make a general statement about the accuracy of the MP2 or B3LYP approaches vis-a-vis that of CCSD(T). For example, the BDEs of the N–EH_3 donor-acceptor bonds (E = B, Al, Ga) are very similar at the MP2/II and B3LYP/II levels of theory. They also agree with the CCSD(T)/II value for the N–BH_3 bond. However, for the N–BCl_3 and N–BBr_3 bonds, the B3LYP/II

Table 7.17 Calculated BDEs D_e (D_0) (kcal/mol) of nitrido[a] and phosphido[b] complexes.

Molecule	Bond	MP2/II	B3LYP/II	CCSD(T)/ II[c]
$Cl_2(PH_3)_3ReN(BH_3)$	N–BH_3	31.9 (29.3)	33.1 (30.4)	31.8 (29.2)
$Cl_2(PH_3)_3ReN(BCl_3)$	N–BCl_3	32.6 (31.2)	23.8 (22.4)	30.2 (28.8)
$Cl_2(PH_3)_3ReN(BBr_3)$	N–BBr_3	36.2 (35.0)	25.3 (24.1)	31.6 (30.4)
$Cl_2(PH_3)_3ReN(AlH_3)$	N–AlH_3	27.1 (25.3)	25.4 (23.6)	
$Cl_2(PH_3)_3ReN(AlCl_3)$	N–$AlCl_3$	43.7 (42.6)	36.8 (35.7)	
$Cl_2(PH_3)_3ReN(AlBr_3)$	N–$AlBr_3$	42.0 (41.0)	33.9 (32.9)	
$Cl_2(PH_3)_3ReN(GaH_3)$	N–GaH_3	18.5 (17.0)	14.9 (13.4)	
$Cl_2(PH_3)_3ReN(GaCl_3)$	N–$GaCl_3$	35.0 (33.5)	28.3 (26.8)	
$Cl_2(PH_3)_3ReN(GaBr_3)$	N–$GaBr_3$	33.2 (31.9)	23.3 (22.0)	
$Cl_2(PH_3)_3ReN(NO)$	N–O	100.2 (97.6)	101.2 (98.6)	
$Cl_2(PH_3)_3ReN(NS)$	N–S	66.5 (65.0)	65.1 (63.6)	
$Cl_2(PH_3)_3ReN(NSe)$	N–Se	47.3 (46.4)	47.2 (46.3)	
$Cl_2(PH_3)_3ReN(NTe)$	N–Te	35.2 (34.7)	36.3 (35.8)	
$Mo(PS)(NH_2)_3$	P–S	34.6	57.9	47.6
$W(PS)(NH_2)_3$	P–S	37.4	50.3	42.2
	Mo–NH_3	10.3	7.2	15.8
$W(P)(NH_2)_3(NH_3)$	W–NH_3	16.0	8.9	17.4
$Mo(PS)(NH_2)_3(NH_3)$	P–S	53.9	71.6	62.9
	Mo–NH_3	29.6	22.8	31.2
$W(PS)(NH_2)_3(NH_3)$	P–S	56.0	66.4	58.5
	W–NH_3	32.4	24.9	33.7

[a] BDEs taken from Ref. 86.
[b] BDEs taken from Ref. 87.
[c] At B3LYP/II optimized geometries.

level of theory produces significantly lower BDEs than the MP2/II one. Since the CCSD(T)/II values for the N–BCl_3 and N–BBr_3 BDEs coincide with the MP2/II estimates, we believe that for the N–EX_3 bonds the latter are more reliable than their B3LYP/II counterparts. On the other hand, the two methods yield nearly identical BDEs for the N–X bonds, where X is a chalcogen.

The results for the P–S BDEs in the phosphido complexes are even more confusing. The B3LYP/II predictions are significantly larger than the MP2/II ones, while the CCSD(T)/II level of theory produces intermediate values of BDEs. Moreover, it has been found that the MP2/II BDEs for the molybdenum complexes are lower than those for the tungsten complexes, while the predictions of the B3LYP/II and CCSD(T)/II levels of theory are that the BDEs of the MoP–S bonds are significantly higher than those of the WP–S bonds. The MP2/II BDEs of the Mo–

NH_3 and W–NH_3 donor-acceptor bonds appear to be more reliable as they agree quite well with the CCSD(T)/II estimates.

14. MAIN GROUP COMPLEXES OF GROUP-13 LEWIS ACIDS EX_3 (E= B - Tl; X = H, F, Cl)

The nature of the donor-acceptor interactions in main group complexes has also been studied theoretically [88-91]. The calculated BDEs of main-group complexes predicted at the MP2/II level of theory are in very good agreement with experimental results [88]. This is an important difference from the performance of the MP2/II level of theory for TM complexes, where the computed BDEs are always too high. Table 7.18 lists the calculated BDEs for complexes of group-13 Lewis acids EX_3 with various Lewis bases.

A comparison of the calculated and experimental data indicates that the MP2 values obtained with the basis sets II do not differ significantly from those afforded by the larger basis set TZP. The theoretical values agree very well with the experimental data. The only larger deviation from experimental data has been found for Cl_3B–NMe_3 (Table 7.18). However, a critical examination of the experimental value of 30.5 kcal/mol led us to suggest that it is probably too low [88].

On the basis of the calculated BDEs, it possible to estimate the strength of the Lewis acidity of the EX_3 compounds. The heavier group-13 trifluorides and trichlorides of Al, Ga, and In exhibit similar Lewis acidities and they are generally stronger Lewis acids than BH_3, BF_3, and BCl_3. However, the bond strengths of donor-acceptor complexes depend also on the nature of the Lewis base. For example, the MP2/TZP BDE of Cl_3B–CO is only 2.3 kcal/mol, that is markedly less than the F_3Al–CO BDE of 15.1 kcal/mol, while the Cl_3B–$C(NH_2)_3$ BDE of 59.7 kcal/mol is slightly higher than that of F_3Al–$C(NH_2)_2$, namely 58.0 kcal/mol. However, the Lewis acid strengths of boron compounds are generally lower than those of their heavier analogues. The calculations indicate that the Lewis acidity of the former species exhibits the trend $BH_3 > BCl_3 > BF_3$.

Table 7.18 Calculated BDEs D_e (D_0) (kcal/mol) of main group complexes with group-13 Lewis acids EX_3 (E = B, Al, Ga, In; X = H, F, Cl).[a]

Compound	MP2/II	MP2/TZP	Exp.
H_3B-H_2	2.1 (-3.1)	4.6 (-0.6)	
$H_3B-C_2H_4$	13.0 (9.0)	14.3 (10.3)	
$H_3B-C_2H_2$	7.4 (4.2)	7.9 (4.6)	
H_3B-CO	25.6 (22.2)	26.4 (23.0)	24.6[b]
H_3B-NH_3	34.6 (29.2)	33.7 (28.3)	31.1[b]
H_3B-NMe_3	41.3 (36.4)	43.6 (38.7)	38.3[b]
H_3B-CS	43.0 (40.2)		
H_3B-PF_3	25.8 (22.5)		
H_3B-PCl_3	20.9 (17.9)		
H_3B-PMe_3	38.9 (35.5)		
F_3B-H_2	0.6 (-0.3)	0.7 (0.2)	
$F_3B-C_2H_4$	4.6 (3.8)	3.6 (2.8)	
$F_3B-C_2H_2$	4.3 (3.8)	3.3 (2.7)	
F_3B-CO	4.0 (3.4)	3.2 (2.5)	
F_3B-NH_3	26.8 (23.5)	23.5 (19.9)	
F_3B-NMe_3	36.1 (33.3)	33.3 (30.5)	31.0±1.1[d]
F_3B-CH_3CN	8.0 (7.5)	7.2 (6.7)	12.0±0.8[d]
$F_3B-PhCHO$	14.3 (13.0)	11.9 (10.6)	15.5±1.0[d]
F_3B-MA[c]	14.3 (12.9)	11.8 (10.4)	
F_3B-HCN	6.6 (5.9)	5.8 (5.1)	
F_3B-Me_2O	19.3 (17.5)	16.7 (14.9)	17.6±0.8[d]
F_3B-CS	10.6 (9.4)		
F_3B-PF_3	2.4 (2.0)		
Cl_3B-H_2	0.2 (0.2)	0.5 (0.1)	
$Cl_3B-C_2H_4$	2.4 (1.9)	3.1 (2.6)	
$Cl_3B-C_2H_2$	2.1 (1.8)	2.6 (2.3)	
Cl_3B-CO	2.0 (1.7)	2.3 (1.8)	
Cl_3B-NH_3	31.4 (27.4)	32.0 (27.8)	
$Cl_3B-C(NH_2)_2$		59.7 (56.7)	
Cl_3B-NMe_3	36.7 (33.6)	41.3 (38.1)	30.5[d]
Cl_3B-CH_3CN	4.4 (4.1)	4.3 (4.0)	
Cl_3B-CS	4.1 (3.7)		

Table 7.18 (continued).

Compound	MP2/II	MP2/TZP	Exp.
$Cl_3B–PH_3$	2.2 (1.6)		
$Cl_3B–PF_3$	-1.3 (-2.6)		
$Cl_3B–PMe_3$	31.5		
$Cl_3Al–CO$		14.2 (12.8)	
$Cl_3Al–NH_3$		44.1 (41.0)	
$Cl_3Al–C(NH_2)_2$		61.6 (59.1)	
$Cl_3Al–EtCClO$	24.9 (24.0)	23.3 (22.4)	
$Cl_3Al–NMe_3$	50.2 (47.6)	49.5 (46.9)	47.5±2.0[e]
$F_3Al–CO$		15.1 (13.7)	
$F_3Al–NH_3$		43.4 (40.4)	
$F_3Al–C(NH_2)_2$		58.0 (55.5)	
$Cl_3Ga–CO$		9.7 (8.6)	
$Cl_3Ga–NH_3$		35.4 (32.6)	
$Cl_3Ga–C(NH_2)_2$		57.5 (55.1)	
$F_3Ga–CO$		12.5 (11.3)	
$F_3Ga–NH_3$		38.5 (35.6)	
$F_3Ga–C(NH_2)_2$		59.5 (56.8)	
$Cl_3In–CO$		12.0 (10.9)	
$Cl_3In–NH_3$		38.7 (36.0)	
$Cl_3In–C(NH_2)_2$		57.6 (55.4)	
$F_3In–CO$		14.2 (13.0)	
$F_3In–NH_3$		40.9 (38.2)	
$F_3In–C(NH_2)_2$		60.1 (57.5)	

[a] BDEs taken from Refs. 88-91.

[b] Ref. 85.

[c] MA = malonaldehyde.

[d] Experimental values [92] corrected by 2.4±0.7 kcal/mol for solvent effects.

[e] Ref. 93.

15. MAIN GROUP COMPLEXES OF BeO

To investigate complexes of the Lewis acid BeO may seem strange from the experimental point of view. BeO is a polymeric solid with a high melting point and it is very difficult to obtain monomeric BeO. Moreover, beryllium is very poisonous and its compounds are difficult

Table 7.19 Calculated BDEs D_e (D_0) (kcal/mol) of main group complexes with the Lewis acid BeO.[a]

Molecule	MP2/ 6-31G(d,p)	MP4/ 6-311G(2df,2pd)
HeBeO	4.7 (3.3)	4.6 (3.3)
NeBeO	10.7 (9.7)	8.9 (7.9)
ArBeO	12.9 (11.9)	17.1 (16.1)
KrBeO	15.5 (14.4)	12.9 (11.8)
XeBeO	18.5 (17.4)	15.8 (14.7)
$N_2BeO(\pi)$	10.2 (9.3)	11.2 (10.3)
$N_2BeO(\sigma)$	28.3 (26.0)	32.3 (30.0)
OCBeO	43.0 (40.5)	43.3 (40.8)
COBeO	21.6 (19.6)	20.4 (18.4)
C_2H_2BeO	41.9 (40.2)	41.3 (39.6)
C_2H_4BeO	44.1 (41.5)	45.2 (42.6)
H_2BeO	18.3 (14.7)	18.5 (14.9)
H_3NBeO	65.8 (62.0)	65.7 (62.3)
Me_3NBeO	69.5 (65.9)	69.5 (66.4)
H_2COBeO	47.8 (44.7)	47.1 (44.2)
O_2BeO	20.6 (18.8)	24.3 (22.7)

[a] BDEs taken from Refs. 95, 97, and 98.

to handle. Indeed, the research on chemistry of the BeO complexes was prompted by purely theoretical investigations. During the search for stable compounds of light noble gas elements, it has been found that BeO should be an unusually strong Lewis acid forming relatively strong bonds with the very weak Lewis bases He, Ne and Ar [94, 95]. The complexes of BeO with other Lewis bases L have been subsequently examined [96-101]. The analysis of the bonding interactions and the calculated BDEs of the OBe–L bonds have shown that BeO could be the strongest neutral Lewis acid of all main-group compounds [96]. Experimental support of this theoretical prediction came later when Andrews and coworkers reported on pulsed laser ablation experiments on beryllium in the presence of noble gases and other weak Lewis bases [102-104]. These workers identified the noble gas complexes NgBeO (Ng = Ar, Kr, Xe), the carbonyl complexes OC–BeO, and the oxygen-bonded isomer CO–BeO that had been predicted theoretically [95, 96]. Although the chemistry of BeO complexes may be considered of little interest from a synthetic standpoint, it is relevant to the understanding of the strength and nature of main-group donor-acceptor complexes.

Table 7.19 lists the BDEs for BeO complexes predicted at the MP2/6-31G(d,p) level of theory, which is equivalent to MP2/II, and at the MP4(SDTQ) level with the large TZ2P-quality 6-311G(2dp,2df) valence basis set. The latter calculations were performed in order to obtain quantitatively reliable BDEs. For some of the less strongly bonded complexes, the size of the BSSE has been estimated via the counterpoise correction. The calculated bond energies given in Table 7.19 clearly indicate that BeO is indeed a stronger Lewis acid than the group-13 compounds EX_3 (Table 7.18). BeO binds to helium with a BDE of ca. 2 - 3 kcal/mol. This is a much stronger bond than those occurring in normal van der Waals complexes of helium that have typically a bond energy of < 20 cm^{-1} (< 0.06 kcal/mol). The complex OBe–NMe_3 has a theoretically predicted BDE of 69.5 kcal/mol. It is the most strongly bonded main-group complex examined theoretically so far. This theoretical prediction still awaits experimental verification!

16. CONCLUSION

The compilation of calculated bond dissociation energies of transition metal compounds and main group complexes, obtained at a standard level of theory that is not very expensive computationally, shows that quantum chemistry can provide important thermodynamic data for a wide range of compounds. The theoretical results are quite reliable and can be obtained with much less effort than it is incurred in the course of experimental work. Nevertheless, it is very important to emphasize that an indiscriminate and uncritical use of theoretical methods to calculate BDEs and other properties of molecules should be strongly discouraged. Although the theoretical approaches and software are equally sophisticated, quantum-chemical programs are unfortunately not as sensitive as experimental techniques to unqualified usage. It requires experience and insight to interpret the calculated data in a critical and correct way. Thus, although we encourage experimentalists to employ modern quantum chemical tools to complement their research, we strongly suggest that an experienced theoretician should be consulted before any conclusions about the meaning of the theoretical results are drawn.

ACKNOWLEDGEMENTS

GF thanks his previous and present coworkers who contributed to the calculations yielding the data that made this compilation possible. We also thank Inga Ganzer for her great efforts to compile the tables. This work was financially supported by the Deutsche Forschungsgemeinschaft and the Fonds der Chemischen Industrie. Excellent service by the Hochschulrechenzentrum of the Philipps-Universitt Marburg is gratefully acknowledged. Additional computer time was provided by the HLRZ Stuttgart and the HHLRZ Darmstadt.

REFERENCES

1. P. v. R. Schleyer, N. L. Allinger, T. Clark, P. A. Kollman, H. F. Schaefer III, and P. R. Scheiner (Eds.), *Encyclopedia of Computational Chemistry*, Vols. 1-5, Wiley-VCH, Chichester (1998).
2. For applications of quantum chemical methods on transition metal compounds see the articles which appeared in the special issue on *Computational Transition Metal Chemistry* in *Chem. Rev.* **100** (2000).
3. G. Frenking and N. Fröhlich, *Chem. Rev.* **100**, 717 (2000).
4. G. Frenking and U. Pidun, *J. Chem. Soc. Dalton Trans.* 1653 (1997).
5. N. Fröhlich and G. Frenking, *Solid State Organometallic Chemistry: Methods and Application*, M. Gielen and B. Wrackmeyer (Eds.), Wiley-VCH, New York (1999) p.173.
6. C. Boehme, J. Uddin, and G. Frenking, *Coord. Chem. Rev.* **197**, 249 (2000).
7. A. J. Lupinetti, S. H. Strauss, and G. Frenking, *Prog. Inorg. Chem.*, in press.
8. M. Torrent, M. Sola, and G. Frenking, *Chem. Rev.* **100**, 439 (2000).
9. V. Jonas, G. Frenking, and M. T. Reetz, *J. Comput. Chem.* **13**, 919 (1992).
10. V. Jonas, Ph.D. thesis, Marburg (1993).
11. P. J. Hay and W. R. Wadt, *J. Chem. Phys.* **82**, 299 (1985).
12. R. Ditchfield, W. J. Hehre, and J. A. Pople, *J. Chem. Phys.* **54**, 724 (1971).
13. W. J. Hehre, R. Ditchfield, and J. A. Pople, *J. Chem. Phys.* **56**, 2257 (1972).
14. P. C. Hariharan and J. A. Pople, *Mol. Phys.* **27**, 209 (1974).
15. P. C. Hariharan and J. A. Pople, *Theor. Chim. Acta* **28**, 213 (1973).
16. M. S. Gordon, *Chem. Phys. Lett.* **76**, 163 (1980).
17. M. Dolg, U. Wedig, H. Stoll, and H. Preuss, *J. Chem. Phys.* **86**, 866 (1987).
18. G. Frenking, I. Antes, M. Boehme, S. Dapprich, A. W. Ehlers, V. Jonas, A. Neuhaus, M. Otto, R. Stegmann, A. Veldkamp, and S. F. Vyboishchikov, *Reviews in Computational Chemistry*, Vol. 8, K. B. Lipkowitz and D. B. Boyd (Eds.), VCH, New York (1996), p. 63.

19. S. Dapprich, U. Pidun, A. W. Ehlers, and G. Frenking, *Chem. Phys. Lett.* **242**, 521 (1995).

20. A. D. Becke, *Phys. Rev.* **A 38**, 3098 (1988); J. P. Perdew, *Phys. Rev.* **B 33**, 8822 (1986).

21. A. D. Becke, *J. Chem. Phys.* **98**, 1372 (1993); A. D. Becke, *J. Chem. Phys.* **98**, 5648 (1993); C. Lee, W. Yang, and R. G. Parr, *Phys. Rev.* **B 37**, 785 (1988); P. J. Stephens, F. J. Devlin, C. F. Chabalowski, and M. J. Frisch, *J. Phys. Chem.* **88**, 11623 (1994).

22. M. Diedenhofen, T. Wagener, and G. Frenking, *Computational Organometallic Chemistry* Marcell Dekker, New York, in print.

23. D. Andrae, U. Häussermann, M. Dolg, and H. Stoll, *Theor. Chim. Acta* **77**, 123 (1990).

24. P. Pyykkö, *Chem. Rev.* **88**, 563 (1988).

25. J. Almlöf and O. Gropen, *Rev. Comput. Chem.* **8**, 203 (1996).

26. P. Schwerdtfeger and M. Seth, *Encyclopedia of Computational Chemistry*, P. v. R. Schleyer, N. L. Allinger, T. Clark, P. A. Kollman, H. F. Schaefer III, P. R. Scheiner (Eds.), Vol. 4, Wiley-VCH, Chichester (1998), p. 2480.

27. L. A. Barnes, M. Rosi, and C. W. Bauschlicher, *J. Chem. Phys.* **83**, 609 (1990).

28. I. Antes and G. Frenking, *Organometallics* **14**, 4263 (1995).

29. Gaussian 94: M. J. Frisch, G. W. Trucks, H. B. Schlegel, P. M. W. Gill, B. G. Johnson, M. A. Robb, J. R. Cheeseman, T. A. Keith, G. A. Petersson, J. A. Montgomery, K. Raghavachari, M. A. Al-Laham, V. G. Zakrzewski, J. V. Ortiz, J. B. Foresman, J. Cioslowski, B. B. Stefanov, A. Nanayakkara, M. Challacombe, C. Y. Peng, P. Y. Ayala, W. Chen, M. W. Wong, J. L. Andres, E. S. Replogle, R. Gomberts, R. L. Martin, D. J. Fox, J. S. Binkley, D. J. Defrees, I. Baker, J. J. P. Stewart, M. Head-Gordon, C. Gonzalez, and J. A. Pople, Gaussian Inc., Pittsburgh, PA (1995).

30. Gaussian 98: M. J. Frisch, G. W. Trucks, H. B. Schlegel, G. E. Scuseria, M. A. Robb, J.R. Cheeseman, V. G. Zakrzewski, J. A. Montgomery, R. E. Stratmann, J. E. Burant, S. Dapprich. J. M. Milliam, A. D. Daniels, K. N. Kudin, M. C. Strain, O. Farkas, J. Tomasi, V. Barone, M. Cossi, R. Cammi, B. Mennucci, C. Pomelli, C. Adamo, S. Clifford, J. Ochterski, G. A. Petersson, P. Y. Ayala, Q. Cui, K. Morokuma, D. K. Malick, A. D. Rabuck, K. Raghavachari, J. B. Foresman, J. Cioslowski, J. V. Ortiz, B.. B. Stefanov, G. Liu, A. Liashenko, P. Piskorz, I. Komaromi, R. Gomberts, R. L. Martin, D. J. Fox, T. A. Keith, M. A. Al-Laham, C. Y. Peng, A. Nanayakkara, C. Gonzalez, M. Challacombe, P. M. W. Gill, B. G. Johnson, W. Chen, M. W. Wong, J. L. Andres, M. Head-Gordon, E. S. Replogle, and J. A. Pople, Gaussian Inc., Pittsburgh, PA (1998).

31. R. Ahlrichs, *Encyclopedia of Computational Chemistry*, P. v. R. Schleyer, N. L. Allinger, T. Clark, P. A. Kollman, H. F. Schaefer III, P. R. Scheiner (Eds.), Vol. 5, Wiley-VCH, Chichester (1998), p. 3123.

32. ACES II, an ab initio program system written by J. F. Stanton, J. Gauss, J. D. Watts, W. J. Lauderdale, and R. J. Bartlett, University of Florida, Gainesville, FL (1991).

33. H. J. Werner and P. J. Knowles, MOLPRO, University of Sussex, Sussex, UK.

34. E. J. Baerends, D. E. Ellis, and P. Ros, *Chem. Phys.* **2**, 41 (1973); G. teVelde and E. J. Baerends, *J. Comput. Phys.* **99**, 84 (1992).
35. T. Ziegler, V. Tschinke, and E. J. Baerends, *J. Chem. Phys.* **74**, 1271 (1981); T. Ziegler, V. Tschinke, E .J. Baerends, J. G. Snijders, and W. Ravenek, *J. Phys. Chem.* **93**, 3050 (1989).
36. E. van Lenthe, E. J. Baerends, and J. G. Snijders, *J. Chem. Phys.* **99**, 4597 (1993); E. van Lenthe, E. J. Baerends, and J. G. Snijders, *J. Chem. Phys.* **101**, 9783 (1994).
37. A. W. Ehlers and G. Frenking, *Organometallics* **14**, 423 (1995).
38. A. W. Ehlers and G. Frenking, *J. Am. Chem. Soc.* **116**, 1514 (1994).
39. A. Diefenbach, M. Bickelhaupt, and G. Frenking, *J. Am. Chem. Soc.* **122**, 6449 (2000).
40. R. K. Szilagyi and G. Frenking, *Organometallics* **16**, 4807 (1997).
41. M. Dörr, diploma thesis, Marburg (2001).
42. A. W. Ehlers, Ph.D. thesis, Marburg (1994).
43. A. E. Stevens, C. S. Feigerle, and W. C. Lineberger, *J. Am. Chem. Soc.* **104**, 5026 (1982).
44. K. E. Lewis, D. M. Golden, and G. P. Smith, *J. Am. Chem. Soc.* **106**, 3905 (1984).
45. R. Huq, A.J. Poë, and S. Chawla, *Inorg. Chim. Acta* **38**, 121 (1980).
46. J.-K. Shen, Y.-C. Gao, Q.-Z. Shi, and F. Basolo, *Inorg. Chem.* **28**, 4304 (1989).
47. A. J. Lupinetti, V. Jonas, W. Thiel, S. H. Strauss, and G. Frenking, *Chem. Eur. J.* **5**, 9 (1999).
48. F. Meyer, Y. M. Chen, and P. Armentrout, *J. Am. Chem. Soc.* **117**, 4071 (1995).
49. A. W. Ehlers, S. Dapprich, S. F. Vyboishchikov, and G. Frenking, *Organometallics* **15**, 105 (1996).
50. S. Dapprich, U. Pidun, A. W. Ehlers, and G. Frenking, *Chem. Phys. Lett.* **242**, 521 (1995).
51. S. Dapprich and G. Frenking, *Angew. Chem.* **107**, 383 (1995); S. Dapprich and G. Frenking, *Angew. Chem. Int. Ed. Engl.* **34**, 354 (1995).
52. G. D. Michels, G. D. Flesh, and H. J. Svec, *Inorg. Chem.* **19**, 479 (1980).
53. J. R. Wells, P. G. House, and E. J. Weitz, *Phys. Chem.* **98**, 8343 (1994).
54. U. Pidun and G. Frenking, *Organometallics* **14**, 5325 (1995).
55. A. Kovács and G. Frenking, *Organometallics*, in press.
56. I. W. Stolz, G. R. Dobson, and R. K. Sheline, *Inorg. Chem.* **2**, 1264 (1963).
57. Y. Zheng, W. Wong, J. Lin, Y. She, and K. Fu, *Chem. Phys. Lett.* **202**, 148 (1993).
58. Y. Chen, M. Hartmann, and G. Frenking, *Z. Anorg. Allg. Chem.*, in press.
59. J. J. Turner, M. B. Simpson, M. Poliakoff, and W. B. Maier II, *J. Am. Chem. Soc.* **105**, 3898 (1983).
60. K. Wichmann, Diploma thesis, Marburg (2000); K. Wichmann and G. Frenking, to be published.
61. A. W. Ehlers, G. Frenking, and E. J. Baerends, *Organometallics* **16**, 4896 (1997).

62. J. R. Wells and E. Weitz, *J. Am. Chem. Soc.* **114**, 2783 (1992).
63. B. H. Weiller, *J. Am. Chem. Soc.* **114**, 10910 (1992).
64. K. H. Dötz, H. Fischer, P. Hofmann, F. R. Kreissl, U. Schubert, and K. Weiss, *Transition Metal Carbene Complexes*, Verlag Chemie, Weinheim (1983); R. R. Schrock, *Acc. Chem. Res.* **12**, 98 (1979).
65. S. F. Vyboishchikov and G. Frenking, *Chem. Eur. J.* **4**, 1428 (1998).
66. C. Boehme and G. Frenking, *Organometallics* **17**, 5801 (1998).
67. S. F. Vyboishchikov and G. Frenking, *Chem. Eur. J.* **4**, 1439 (1998).
68. M. Böhme, T. Wagener, and G. Frenking, *J. Organomet. Chem.* **520**, 31 (1996).
69. J. Uddin, S. Dapprich, G. Frenking, and B. Yates, *Organometallics* **18**, 457 (1999).
70. A. Kovács and G. Frenking, *Organometallics* **18**, 887 (1999).
71. U. Pidun and G. Frenking, *J. Organomet. Chem.* **525**, 269 (1996).
72. J. Wei, D. Stetzkamp, B. Nuber, R. A. Fischer, C. Boehme, and G. Frenking, *Angew. Chem.* **109**, 95 (1997); J. Wei, D. Stetzkamp, B. Nuber, R. A. Fischer, C. Boehme, and G. Frenking, *Angew. Chem. Int. Ed. Engl.* **36**, 70 (1997).
73. R. A. Fischer, M. M. Schulte, J. Weiss, L. Zsolnai, A. Jacobi, G. Huttner, G. Frenking, C.Boehme, and S. F. Vyboishchikov, *J. Am. Chem. Soc.* **120**, 1237 (1998).
74. C. Boehme and G. Frenking, *Chem. Eur. J.* **5**, 2184 (1999).
75. W. Uhl, M. Benter, S. Melle, W. Saak, G. Frenking, and J. Uddin, *Organometallics* **18**, 3778 (1999).
76. J. Uddin, C. Boehme, and G. Frenking, *Organometallics* **19**, 571 (2000).
77. M. Esser, B. Neumüller, W. Petz, J. Uddin, and G. Frenking, *Z. Anorg. Allg. Chem.* **626**, 915 (2000).
78. D. Weiss, T. Steinke, M. Winter, R.A. Fischer, N. Fröhlich, J. Uddin, and G. Frenking, *Organometallics* **19**, 4583 (2000).
79. J. Uddin and G. Frenking, *J. Am. Chem. Soc.* **123**, 1683 (2001).
80. C. Boehme, PhD thesis, Marburg (1998).
81. R. A. Fischer, A. Miehr, H. Hoffmann, W. Rogge, C. Boehme, G. Frenking, and E. Herdtweck, *Z. Anorg. Allg. Chem.* **625**, 1466 (1999).
82. K. T. Giju, F. M. Bickelhaupt, and G. Frenking, *Inorg. Chem.* **39**, 4776 (2000).
83. G. A. Irvine, M. J. G. Lesley, T. B. Marder, N. C. Norman, C. R. Rice, E. G. Robins, W. R. Roper, G. R. Whittell, and L. J. Wright, *Chem. Rev.* **98**, 2685 (1998).
84. P. B. Armentrout and B. L. Kickel, *Organometallic Ion Chemistry* S. B. Freiser (Ed.), Kluwer, Dordrecht, in press; P. B. Armentrout, personal communication (1994).
85. S. G. Lias, J. E. Bartmess, J. F. Liebman, J.L. Holmes, R. D. Levin, and W. G. Mallard, *J. Phys. Chem. Ref. Data* **17**, Suppl.1 (1998).
86. S. F. Vyboichshikov and G. Frenking, *Theor. Chem. Acc.* **102**, 300 (1999).
87. T. Wagener and G. Frenking, *Inorg. Chem.* **37**, 1805 (1998).
88. V. Jonas, G. Frenking, and M. T. Reetz, *J. Am. Chem. Soc.* **116**, 8741 (1994).

89. S. Fau and G. Frenking, *Mol. Phys.* **96**, 519 (1999).
90. A. Beste, O. Krämer, A. Fischer, and G. Frenking, *Eur. J. Inorg. Chem.* 2037 (1999).
91. A. Diefenbach and G. Frenking, to be published.
92. J. -F. Gal and P. -C. Maria, *Progress in Phys. Org. Chem.* **17**, 159 (1990).
93. G. A. Andersen, F. R. Forgaard, and A. Haaland, *Acta Chem. Scand.* **26**, 1947 (1972).
94. W. Koch, J. R. Collins, and G. Frenking, *Chem. Phys. Lett.* **132**, 330 (1986).
95. G. Frenking, W. Koch, J. Gauss, and D. Cremer, *J. Am. Chem. Soc.* **110**, 8007 (1988).
96. W. Koch and G. Frenking, *Molecules in Natural Science and Medicine. An Encomium for Linus Pauling*, Z. B. Maksic and M. Eckert-Maksic (Eds.), Ellis Horwood, New York (1991), p. 225.
97. A. Veldkamp and G. Frenking, *Chem. Phys. Lett.* **226**, 11 (1994).
98. G. Frenking, S. Dapprich, K. F. Köhler, W. Koch, and J. R. Collins, *Mol. Phys.* **89**, 1245 (1996).
99. G. Frenking, W. Koch, and J. R. Collins, *J. Chem. Soc. Chem. Commun.* 1147 (1988).
100. G. Frenking and D. Cremer, *Structure and Bonding*, Vol. 73, Springer Verlag, Heidelberg (1990), p. 17.
101. G. Frenking, W. Koch, F. Reichel, and D. Cremer, *J. Am. Chem. Soc.* **112**, 4240 (1990).
102. C. A. Thompson and L. Andrews, *J. Am. Chem. Soc.* **116**, 423 (1994).
103. L. Andrews and T. J. Tague, *J. Am. Chem. Soc.* **116**, 6856 (1994).
104. C. A. Thompson and L. Andrews, *J. Chem. Phys.* **100**, 8689 (1994).

Chapter 8

Semiempirical Thermochemistry: A Brief Survey

Walter Thiel
Max-Planck-Institut für Kohlenforschung, Kaiser-Wilhelm-Platz 1, D-45470 Mülheim, Germany

1. INTRODUCTION

The semiempirical molecular orbital (MO) methods of quantum chemistry [1-12] are widely used in computational studies of large molecules. A number of such methods are available for calculating thermochemical properties of ground state molecules in the gas phase, including MNDO [13], MNDOC [14], MNDO/d [15-18], AM1 [19], PM3 [20], SAM1 [21, 22], OM1 [23], OM2 [24, 25] MINDO/3 [26], SINDO1 [27, 28], and MSINDO [29-31]. MNDO, AM1, and PM3 are widely distributed in a number of software packages, and they are probably the most popular semiempirical methods for thermochemical calculations. We shall therefore concentrate on these methods, but shall also address other NDDO-based approaches with orthogonalization corrections [23-25].

The semiempirical calculation of thermochemical properties has been reviewed recently [32]. The present chapter is a condensed and updated version of this previous review. It outlines the theoretical background of semiempirical methods, defines specific conventions, provides statistical evaluations, and discusses the performance with regard to thermochemical properties.

J. Cioslowski (ed.), Quantum-Mechanical Prediction of Thermochemical Data, 235–245.

2. THEORETICAL BACKGROUND

Most current general-purpose semiempirical methods are based on molecular orbital (MO) theory and employ a minimal basis set for the valence electrons; electron correlation is treated explicitly only if this is necessary for the appropriate zeroth-order description. Compared with the ab initio MO formalism, many of the less important (many-center) integrals are neglected to speed up the calculations; traditionally there are three levels of integral approximation called CNDO, INDO, and NDDO [2, 33], the latter being the least severe one. In an attempt to compensate for the errors introduced by these simplifications, the remaining integrals are represented by suitable parametric expressions and calibrated against reliable experimental or accurate theoretical reference data. The quality of semiempirical results will thus depend on both the chosen theoretical model and the parameterization.

MNDO, AM1, and PM3 are based on the same semiempirical model [12, 13], and differ only in minor details of the implementation of the core-core repulsions. Their parameterization has focused mainly on heats of formation and geometries, with the use of ionization potentials and dipole moments as additional reference data. Given the larger number of adjustable parameters and the greater effort spent on their development, AM1 and PM3 may be regarded as methods which attempt to explore the limits of the MNDO model through careful and extensive parameterization.

MNDO, AM1, and PM3 employ an *sp* basis without *d* orbitals [13, 19, 20]. Hence, they cannot be applied to most transition metal compounds, and difficulties are expected for hypervalent compounds of main-group elements where the importance of *d* orbitals for quantitative accuracy is well documented at the ab initio level [34]. To overcome these limitations, the MNDO formalism has been extended to *d* orbitals. The resulting MNDO/d approach [15-18] retains all the essential features of the MNDO model.

Due to the integral approximations used in the MNDO model, closed-shell Pauli exchange repulsions are not represented in the Hamiltonian, but are only included indirectly, e.g., through the effective atom-pair correction terms to the core-core repulsions [12]. To account for Pauli repulsions more properly, the NDDO-based OM1 and OM2 methods [23-25] incorporate orthogonalization terms into the one-center or the one- and two-center one-electron matrix elements, respectively. Similar correction terms have also been used at the INDO level [27-31] and probably contribute to the success of methods such as MSINDO [29-31].

In ab initio theory, inclusion of electron correlation is essential to reliable prediction of thermochemical properties such as atomization energies [34-36]. On the other hand, density functional theory (DFT) is also rather successful in this regard [35] due to the incorporation of dynamic electron correlation effects during orbital optimization through the use of a suitable exchange-correlation functional. Conceptually, dynamic electron correlation is also built into the semiempirical SCF-MO methods in an average manner, by using effective two-electron interactions that are damped at small and intermediate distances [12]. Closer analysis shows [8, 14] that dynamic correlation effects are relatively small in MNDO-type methods (due to the representation of the two-electron integrals) and also rather uniform (in related molecules). It is therefore not surprising that these effects can be taken into account in an average manner by a parameterization at the SCF level. Hence, MNDO [13] and the explicitly correlated MNDOC approach [14] yield results of similar accuracy for standard closed-shell ground state molecules, and MNDOC offers advantages only in systems with specific correlation effects [8].

These considerations suggest a general answer to the question of what to expect from semiempirical calculations of thermochemical properties. The results can be reliable only to the extent that the relevant physical interactions are included in the semiempirical model and that the formally neglected features can be absorbed by the semiempirical parameterization in an average sense. Whether a given semiempirical method will be useful for quantitative thermochemical predictions can only be established by careful validation against experimental data or high-level ab initio results.

3. SPECIFIC CONVENTIONS

In quantum-chemical calculations, the equilibrium atomization energy D_e of a given molecule is available from its total energy and the electronic energies of the constituent atoms. Inclusion of zero-point vibrational terms provides the ground state atomization energy D_0. Subtracting D_0 from the sum of the experimental enthalpies of formation of the constituent atoms at 0 K then yields the heat of formation at 0 K (ΔH_f^0) which can be converted to the corresponding heat of formation at 298 K (ΔH_f^{298}) by incorporating suitable thermal corrections.

This procedure is normally followed in ab initio studies and could equally well be applied in semiempirical work. However, in MNDO-type methods, heats of formation at 298 K are traditionally derived in a simpler manner [1, 13]. By formally neglecting the zero-point vibrational energies and the thermal corrections between 0 K and 298 K, the heats of formation at 298 K are obtained from the calculated total energies by subtracting the calculated electronic energies of the atoms in the molecule and by adding their experimental heats of formation at 298 K. This procedure implicitly assumes that the zero-point vibrational energies and the thermal corrections are composed of additive increments, which can be absorbed by the semiempirical parameterization. There is some evidence for the approximate validity of this assumption [32].

According to these conventions, the computed total energies for a given molecule differ from the corresponding heats of formation at 298 K by a constant amount which is geometry-independent. In practice, the potential surface from an MNDO-type calculation is normally discussed like a surface from any other quantum-chemical calculation, except that the energies at the minima are translated into heats of formation without explicitly considering zero-point vibrational energies or thermal corrections.

4. STATISTICAL EVALUATIONS

Comparison with experiment constitutes the ultimate test of theoretical calculations. It is of course essential that the evaluation of theoretical methods is done with regard to reliable experimental reference data. Such a reference set has recently been assembled for the validation of Gaussian-2 (G2) theory [35], containing 148 molecules for which reliable experimental heats of formation at 298 K are available (with a target accuracy of at least 1 kcal/mol). This "G2 neutral test set" provides a means for assessing theoretical methods, and detailed evaluations are already available [35] for the G2 method and its variants, for several DFT approaches (e.g. LDA (SVWN), BLYP, BP86 in the usual notation), and for various HF/DFT hybrid methods (e.g. B3LYP, B3P86). Table 8.1 compares the published results for some of these methods [35] and the more recent Gaussian-3 (G3) theory [36] with those for the semiempirical methods [32].

Among the methods considered in Table 8.1, the G3 approach is most accurate, followed by G2 and B3LYP. The semiempirical methods (especially PM3 and MNDO/d) show similar errors as BLYP, whereas BP86 and particularly LDA(SVWN) overbind strongly [35]. To put these results into perspective, it should be noted that the complete geometry optimization for all test molecules combined takes less than 3 seconds of cpu time for MNDO, AM1, or PM3 when using our current program [37] on a Compaq XP1000 workstation. In view of this very

Table 8.1 Mean absolute deviations (kcal/mol) for the molecules of the G2 neutral test set.[a]

Method	All	First-row	Second-row
G2	1.56 (148)	1.53 (93)	1.72 (47)
G3	0.94	0.81	1.24
LDA (SVWN)	91.16	108.46	67.34
BP86	20.19	24.75	13.85
BLYP	7.09	7.38	7.13
B3LYP	3.11	2.42	4.46
MNDO	9.32 (146)	7.71	12.44
AM1	7.81 (142)	7.44	8.86
PM3	7.01 (144)	6.86	7.15
MNDO/d	7.26	7.71	6.10
OM1	b	4.64	b
OM2	b	3.36 (81)	b

[a] Unless noted otherwise, the number of molecules studied that is given in parentheses for G2 also applies to the other methods (see Ref. 32 for more details). The first- and second-row subgroups exclude compounds containing the Li, Be, B or Na atoms that have not yet been parameterized in some of the semiempirical methods. The ab initio and DFT results have been derived from the published data [35,36].

[b] Not yet available due to the lack of parameters.

low computational effort, the performance of the semiempirical methods appears quite acceptable.

The deviations between theory and experiment normally tend to be somewhat smaller for first-row than for second-row compounds. Among the established semiempirical methods, PM3 seems to be the best for the first-row compounds, but the OM1 and OM2 approaches with orthogonalization corrections [23-25] perform even better, with mean absolute deviations being around 3-5 kcal/mol. For the second-row compounds, MNDO/d is currently the most accurate among the semiempirical methods considered. This performance has been attributed [16-18] to the use of an *spd* basis which allows a balanced description of normalvalent and hypervalent molecules. OM1 and OM2 have not yet been parameterized for second-row elements.

Table 8.2 Mean absolute deviations (kcal/mol) for first-row reference molecules.[a]

Method	First-row	CH	CHN	CHNO	CHNOF
MNDO	7.35 (181)	5.81 (58)	6.24 (32)	7.12 (48)	10.50 (43)
AM1	5.80	4.89	4.65	6.79	6.76
PM3	4.71	3.79	5.02	4.04	6.45
OM1	3.87	2.49	4.27	4.12	5.17
OM2	3.07 (138)	1.75	3.92	4.11	[b]

[a] Unless noted otherwise, the number of molecules studied that is given in parentheses for MNDO also applies to the other methods. The subgroups consist of compounds containing the elements indicated.
[b] Not yet available.

In an overall assessment, the established semiempirical methods perform reasonably for the molecules in the G2 neutral test set. With an almost negligible computational effort, they provide heats of formation with typical errors around 7 kcal/mol. The semiempirical OM1 and OM2 approaches that go beyond the MNDO model and are still under development promise an improved accuracy (see Table 8.1).

Even though the G2 neutral test set is very valuable, it is biased towards small molecules and does not cover all bonding situations that may arise for a given element. The validation of semiempirical methods has traditionally been done using larger test sets which, however, have the drawback that the experimental reference data are often less accurate than those in the G2 set.

In our own validation sets, experimental heats of formation are preferentially taken from recognized standard compilations [38-40]. If there are enough experimental data for a given element, we normally only use reference values that are accurate to 2 kcal/mol. If there is a lack of reliable data, we may accept experimental heats of formation with a quoted experimental error of up to 5 kcal/mol. This choice is motivated by the target accuracy of the established semiempirical methods. If experimental data are missing for a small molecule of interest, we consider it legitimate [18] to employ computed heats of formation from high-level ab initio methods as substitutes.

Our primary validation set for first-row compounds is derived from the original MNDO development [13, 41], but has been updated to include new experimental data for the reference molecules. Table 8.2 shows

Table 8.3 Mean absolute deviations (kcal/mol) for compounds containing second-row and heavier elements.[a]

Method	All	Normalvalent	Hypervalent
MNDO	29.2 (488)	11.0 (421)	143.2 (67)
AM1	15.3	8.0	61.3
PM3	10.9 (552)	9.6 (485)	19.9
MNDO/d	5.4 (575)	5.4 (508)	5.4

[a] See Refs. 16-18 for details and more element-specific data. Unless noted otherwise, the number of molecules studied that is given in parentheses for MNDO also applies to the other methods.

statistical evaluations for this set which are consistent with those compiled in Table 8.1. In both cases, the mean absolute deviations decrease in the sequence MNDO > AM1 > PM3 > OM1 > OM2. Generally, the errors are smaller here than in Table 8.1 which is probably due to the fact that our larger set contains a larger portion of "normal" organic molecules without "difficult" bonding characteristics.

The data in Table 8.2 refer almost exclusively to closed-shell molecules. A second validation set for first-row compounds [42] contains 38 radicals and radical cations. The mean absolute errors for these species are higher than those in Table 8.2. They amount to 11.08, 9.73, 9.41, 6.70, and 4.79 kcal/mol for MNDO, AM1, PM3, OM1, and OM2, respectively.

In the course of the MNDO/d development [15-18] we have generated new validation sets for second-row and heavier elements. Those for Na, Mg, Al, Si, P, S, Cl, Br, I, Zn, Cd, and Hg have been published [16-18]. The corresponding statistical evaluations for heats of formation [18] are summarized in Table 8.3. It is obvious that MNDO/d shows by far the smallest errors followed by PM3 and AM1. All four semiempirical methods perform reasonably well for normalvalent compounds, especially when considering that more effort has traditionally been spent on the parameterization of the first-row elements. For hypervalent compounds, however, the errors are huge in MNDO and AM1, and still substantial in PM3, in spite of the determined attempt to reduce these errors in the PM3 parameterization [20]. Therefore it seems likely that the improvements in MNDO/d are due to the use of an *spd* basis set [16-18].

Many other statistical evaluations are available in the literature [1-31] that document the thermochemical results obtained from various semiempirical methods. For a convenient reference, Table 8.4 quotes a

Table 8.4 Mean absolute deviations (kcal/mol) from the literature.[a]

Method	Elements	MAD	Ref.[b]
MNDO	C,H,N,O	6.3 (138)	13
MNDO	C,H,N,O	11.2 (276)	13,20
MNDOC	C,H,N,O	5.3 (64)	14
MNDO/d	Si,P,S,Cl,Br,I	5.1 (404)	18
MNDO/d	Na-Cl,Br,I,Zn,Cd,Hg	5.4 (575)	18
AM1	C,H,N,O	5.5 (138)	19
AM1	C,H,N,O	7.5 (276)	19,20
PM3	C,H,N,O	4.4 (138)	20
PM3	C,H,N,O	5.7 (276)	20
SAM1	C,H,N,O,F	4.4 (285)	21,22
SAM1	Si,P,S,Cl,Br,I	9.3 (404)	21,18
SAM1d	Si,P,S,Cl,Br,I	8.2 (404)	21,18
MINDO/3	C,H,N,O	11.0 (138)	26,13
SINDO1	C,H,N,O	9.49 (64)	30,27
SINDO1	Na,Mg,Al,Si,P,S,Cl	28.89 (167)	30,27
MSINDO	C,H,N,O	5.12 (64)	30
MSINDO	Na,Mg,Al,Si,P,S,Cl	7.07 (167)	30
MSINDO	Ga,Ge,As,Se,Br	5.75 (69)	31

[a] The numbers of molecules studied are given in parentheses.

[b] The first reference cites the method. If given, the second reference contains the quoted value of the mean absolute deviation from a later evaluation.

small selection of the published mean absolute deviations (MADs) between computed and experimental heats of formation. Of course, the corresponding values cannot be compared directly, since they are based on different sets of reference molecules and reference data, but they should provide some indication of the errors that can be expected in such calculations. The reader should consult the original literature for further information [1-31].

5. DISCUSSION

The statistical evaluations of the preceding section indicate that the semiempirical MO methods can predict heats of formation with useful accuracy and at very low computational costs. When comparing with ab initio or DFT methods, the following points [32] should be kept in mind, however:

1. In general, errors tend to be more systematic at a given ab initio or DFT level and may therefore often be taken into account by suitable corrections. Errors in semiempirical calculations are normally less uniform and thus harder to correct.

2. The accuracy of the semiempirical results may be different for different classes of compounds, and there are elements that are more "difficult" than others. Such variations in the accuracy are again less pronounced in high-level ab initio and DFT calculations.

3. Semiempirical methods can only be applied to molecules containing elements that have been parameterized, while ab initio and DFT methods are generally applicable.

4. Semiempirical parameterizations require reliable experimental or theoretical reference data and are impeded by the lack of such data. Such problems do not occur in ab initio or DFT approaches.

In spite of these limitations, there are many areas where the established MNDO-type semiempirical methods can be applied successfully in calculations of thermochemical properties. This suggests that the underlying MNDO model includes the physically relevant interactions so that the parameterization can absorb the errors due to the MNDO approximations in an average sense. However, further improvements are clearly needed, and the inclusion of orthogonalization corrections that account for Pauli exchange repulsion indeed seems to enhance the accuracy of the calculated thermochemical properties (see the OM1 and OM2 results). This supports our belief [12] that a theoretically guided search for better models offers the most promising perspective for general-purpose semiempirical methods with better overall performance.

ACKNOWLEDGEMENTS

The author wishes to thank his coworkers for their contributions, particularly M. Kolb, A. Voityuk, and W. Weber.

REFERENCES

1. M. J. S. Dewar, *The Molecular Orbital Theory of Organic Chemistry*, McGraw-Hill, New York (1969).
2. J. A. Pople and D. L. Beveridge, *Approximate Molecular Orbital Theory*, Academic Press, New York (1970).
3. J. N. Murrell and A. J. Harget, *Semiempirical Self-Consistent-Field Molecular Orbital Theory of Molecules*, Wiley, New York (1972).
4. G. A. Segal (ed.), *Modern Theoretical Chemistry*, Plenum, New York, Vols. 7-8 (1977).
5. M. J. S. Dewar, *Science* **187**, 1037 (1975).
6. K. Jug, *Theor. Chim. Acta* **54**, 263 (1980).
7. M. J. S. Dewar, *J. Phys. Chem.* **89**, 2145 (1985).
8. W. Thiel, *Tetrahedron* **44**, 7393 (1988).
9. J. J. P. Stewart, *J. Comp.-Aided Mol. Design* **4**, 1 (1990).
10. J. J. P. Stewart, in *Reviews in Computational Chemistry*, K.B. Lipkowitz and D.B. Boyd, (eds.), VCH Publishers, New York (1990), Vol. 1, p. 45.
11. M. C. Zerner, in *Reviews in Computational Chemistry*, K. B. Lipkowitz and D.B. Boyd, (eds.), VCH Publishers, New York (1991), Vol. 2, p. 313.
12. W. Thiel, *Adv. Chem. Phys.* **93**, 703 (1996).
13. M. J. S. Dewar and W. Thiel, *J. Am. Chem. Soc.* **99**, 4899 (1977); M. J. S. Dewar and W. Thiel, *J. Am. Chem. Soc.* **99**, 4907 (1977).
14. W. Thiel, *J. Am. Chem. Soc.* **103**, 1413 (1981).
15. W. Thiel and A. A. Voityuk, *Theor. Chim. Acta* **81**, 391 (1992).
16. W. Thiel and A. A. Voityuk, *Int. J. Quant. Chem.* **44**, 807 (1992).
17. W. Thiel and A. A. Voityuk, *J. Mol. Struct.* **313**, 141 (1994).
18. W. Thiel and A. A. Voityuk, *J. Phys. Chem.* **100**, 616 (1996).
19. M. J. S. Dewar, E. Zoebisch, E. F. Healy, and J. J. P. Stewart, *J. Am. Chem. Soc.* **107**, 3902 (1985).
20. J. J. P. Stewart, *J. Comp. Chem.* **10**, 209 (1989); J. J. P. Stewart, *J. Comp. Chem.* **10**, 221 (1989).
21. M. J. S. Dewar, C. Jie, and J. Yu, *Tetrahedron* **49**, 5003 (1993).
22. A. J. Holder, R. D. Dennington, and C. Jie, *Tetrahedron* **50**, 627 (1994).
23. M. Kolb and W. Thiel, *J. Comp. Chem.* **14**, 37 (1993).
24. W. Weber, Ph.D. Dissertation, University of Zurich (1996).
25. W. Weber and W. Thiel, *Theor. Chem. Acc.* **103**, 495 (2000).
26. R. C. Bingham, M. J. S. Dewar, and D. H. Lo, *J. Am. Chem. Soc.* **97**, 1285 (1975).
27. D. N. Nanda and K. Jug, *Theor. Chim. Acta* **57**, 95 (1980).
28. K. Jug, R. Iffert, and J. Schulz, *Int. J. Quantum Chem.* **32**, 265 (1987).
29. B. Ahlswede and K. Jug, *J. Comp. Chem.* **20**, 563 (1999).

30. B. Ahlswede and K. Jug, *J. Comp. Chem.* **20**, 572 (1999).

31. K. Jug, G. Geudtner, and T. Homann, *J. Comp. Chem.* **21**, 974 (2000).

32. W. Thiel, *ACS Symp. Ser.* **677**, 142 (1997).

33. J. A. Pople, D. P. Santry, and G. A. Segal, *J. Chem. Phys.* **43**, S 129 (1965).

34. W. J. Hehre, L. Radom, P. v. R. Schleyer, and J. A. Pople, *Ab Initio Molecular Orbital Theory*, Wiley, New York (1986).

35. L. A. Curtiss, K. Raghavachari, P. C. Redfern, and J. A. Pople, *J. Chem. Phys.* **106**, 1063 (1997).

36. L. A. Curtiss, K. Raghavachari, P. C. Redfern, V. Rassolov, and J.A. Pople, *J. Chem. Phys.* **109**, 7764 (1998).

37. W. Thiel, Program MNDO99, Mülheim (1999).

38. J. B. Pedley, R. D. Naylor, and S. P. Kirby, *Thermochemical Data of Organic Compounds, 2nd ed.*, Chapman Hall, London (1986).

39. M. W. Chase, C. A. Davies, J. R. Downey, D. R. Frurip, R. A. McDonald, and A. N. Syverud, *JANAF Thermochemical Tables, 3rd ed.*, *J. Phys. Chem. Ref. Data* **14** (1985), Suppl. 1.

40. S. G. Lias, J. E. Bartmess, J. F. Liebman, J. L. Holmes, R. D. Levin, and W. G. Mallard, *Gas Phase Ion and Neutral Thermochemistry*, *J. Phys. Chem. Ref. Data* **17** (1988), Suppl. 1.

41. M. J. S. Dewar and H. S. Rzepa, *J. Am. Chem. Soc.* **100**, 58 (1978).

42. D. Higgins, C. Thomson, and W. Thiel, *J. Comp. Chem.* **9**, 702 (1988).

Index

Averaged coupled-pair functional (ACPF), 42, 44, 54-55, 57
Atomization energy, 2, 6-11, 16, 19, 21-26, 32-34, 37-38, 40, 46-47, 52-53, 55, 58-60, 83, 169, 237

Basis set
- 3-21G, 104-107, 109, 124, 162
- 3-21G(*), 102-103
- 6-31G, 124, 149-150
- 6-31G(d), 45, 70-71, 73, 76-77, 79, 81-83, 94, 102, 104-107, 109, 122, 151-152, 162, 164-169, 172-173, 175-176, 178-191, 219
- 6-31G(d,p), 149, 188, 227-228
- 6-31G(2df,p), 71, 82, 84-87
- 6-31+G(d), 71
- 6-311G, 150
- 6-311G(d,p), 102-103, 107-109, 125, 145, 150, 189
- 6-311G(2df,2p), 151-152
- 6-311G(2df,2pd), 227-228
- 6-311+G(d,p), 181-182, 188-190, 192
- 6-311+G(2df,p), 124, 175, 179
- 6-311+G(2df,2pd), 102
- 6-311+G(3df,2p), 89, 93, 162, 164, 176-178, 180-192
- 6-311++G, 150
- 6-311++G(2df,2p), 151-152
- 6-311++G(3df,3pd), 146-150
- Ahlrichs's, 150, 218
- CEP-121G, 150
- CEP-31G, 147, 150
- CEP-4G, 150
- correlation-consistent, 4, 33, 110
  - aug-cc-pVnZ, 33, 162
  - aug-cc-pVDZ, 61, 150
  - aug-cc-pVTZ, 150

aug′-cc-pVnZ, 33
cc-pCVnZ, 4-5, 19, 22, 27, 33, 162
cc-pCVDZ, 11, 19-22
cc-pCVTZ, 11, 19, 21-22
cc-pCVQZ, 9, 11, 17, 19, 21, 23, 25, 26
cc-pCV5Z, 11, 19, 21
cc-pCV6Z, 9, 11, 17-21
cc-pVnZ, 4-5, 17-18, 22, 33, 36, 39, 51, 54, 110-117, 119, 162
cc-pVDZ, 6, 8, 17, 39, 110, 113-114, 116-119, 150, 167
cc-pVTZ, 17, 23, 35, 41, 45, 48-49, 60, 113-114, 116-119, 150, 165-169
cc-pVQZ, 17, 60, 83, 113, 117-119, 146-147, 149-150, 155
cc-pV5Z, 7-8, 17, 19, 35, 59-61, 113, 117, 119
cc-pV6Z, 19, 110, 113, 116-117, 119
cc-pVnZ+1, 33
cc-pVTZ+1, 35, 45-47, 57, 164
cc-pVQZ+1, 34-35
cc-pVnZ+2d1f, 33
cc-pV(n+d)Z, 33
cc-pVTZ2P, 45
DZ, 148-150
DZP, 200
Frenking's II, 200-201, 203-226
G3MP2Large, 73, 164
G3Large, 71-72, 83-84, 87, 164
G3XLarge, 83-84, 87
LANL2DZ, 147-148, 150, 155
LANL2MB, 150
SDB-cc-pVnZ, 61
SHC, 150
Slater-type, 13, 202
TZ, 149-150
TZP, 202-204, 211-212, 216-219, 224-226
TZ2P, 228
TZ(2)P, 210-211
Wachter's, 150
well-tempered, 146-147, 150
Basis set limit, 11, 16, 36, 61, 67, 99-102, 110, 114-115, 124, 166
Bond dissociation energy (BDE), 161, 174-177, 179, 192, 201-228
Brueckner doubles, 140

Complete basis set (CBS) method, 35, 68, 100, 102, 107, 125, 127, 188, 193
- CBS-4, 107, 109
- CBS-4M, 103, 108, 122, 124, 127
- CBS/APNO, 178
- CBS-q, 107
- CBS-Q, 31, 48, 58, 60, 109, 163-164, 172-173, 178, 188-189
- CBS-QB3, 31, 48, 60, 103, 108, 119, 125, 127, 164, 172-173, 178, 181-182, 190
- CBS-QCI/APNO, 103, 107-109, 127
- CBS-RAD, 164, 172-176, 178-191, 193

Coupled-cluster theory, 3, 99
- CCSD, 3-7, 9-11, 17-19, 34, 38-41, 56-57, 60, 101, 107-109
- CCSDT, 3, 6, 8, 26, 56, 165
- CCSDT-1a, 39
- CCSD(T), 7-11, 17, 19-28, 33-35, 38-41, 45, 48-49, 53-54, 56-57, 59-60, 67-68, 77, 99, 103, 109-110, 112, 117-119, 162-164, 167-168, 178, 182, 190, 193, 200-201, 203-216, 221-224
- CCSDTQ, 6-7
- CCSDTQ5, 6
- CCSDTQ56, 6

Configuration interaction
- CISD, 34
- FCI, 5, 13-14, 16, 117
- QCISD, 102-103, 164-165, 167-168, 172-173, 178, 182, 190
- QCISD(T), 67-68, 71-72, 77-78, 102, 104, 162-164, 178, 182, 190

Density functional, 68-69, 88, 91-92, 95, 162-163, 167, 190, 200-202, 237-238, 243
- B3LYP, 33-35, 45-49, 54-61, 77, 82, 84-87, 89-95, 102-103, 125, 162-164, 166-167, 169, 173, 176-178, 180-193, 200-201, 203-208, 215, 218, 220, 222-223, 238-239
- B3P86, 238
- BH&HLYP, 48, 58-59
- BLYP, 88-89, 91-92, 95, 162, 178, 182, 190, 238-239
- BP86, 200-201, 203-206, 210-212, 216-219, 238-239
- HCTH-120, 58
- LDA, 88, 90, 95, 238-239
- mPW1K, 49, 58, 59
- mPW1PW91, 58
- SVWN, 88, 238-239

Dyson orbital, 133-134, 136, 140, 142-145, 156

Effective core potential (ECP), 145-151, 155, 200, 204, 215, 219
    Stuttgart-Dresden, 147-150, 155, 202
Electron affinity, 46, 48-49, 52, 61, 69, 73-74, 80, 86-87, 131, 138-140, 151-155
Electron propagator theory, 131-132, 135, 138-139
    2ph-TDA approximation, 140
    ADC(3) approximation, 140
    NR2 approximation, 140
    OVGF approximation, 139
    P3 approximation, 134-135, 140-142, 145-156
Energy extrapolation, 15-16, 26, 33-36, 38-39, 48, 50-51, 57, 60-61, 67, 99-102, 110-120, 124-125, 164
Enthalpy of formation, 50-51, 60-61, 69, 70, 73-74, 80-81, 84-91, 93, 161, 168-173, 237-238, 240, 242
Enthalpy of reaction, 99, 109, 119, 161, 190-192

Gaussian-n (Gn) theory
    G1, 31
    G2, 31, 48, 59, 60, 70, 73, 75-76, 88-89, 91-92, 94-95, 163-164, 170-171, 178, 188-189, 238-239
    G2(MP2), 76, 188-189
    G2(MP2)-RAD, 188-189
    G2(MP2,SVP), 163-164, 181-182, 190
    G2(MP2,SVP)-RAD, 164, 188-189
    G2-RAD(QCISD), 164, 170, 173, 178
    G3, 31, 48, 60, 68, 70, 72-78, 81-86, 88-95, 163-164, 171, 173, 178, 181, 238-239
    G3//B3LYP, 77
    G3(CCSD), 77
    G3(MP2), 68, 73-74, 76, 79, 85-86, 92-94, 163-164, 171, 173, 181-182, 190
    G3(MP2)-RAD, 164, 171, 173, 175-176, 178-179, 182-191, 193
    G3(MP2)-RAD(p), 188-189
    G3(MP3), 68, 73-74, 76, 79, 85-86, 92-94
    G3-RAD, 164, 171, 173
    G3S, 68, 77-81, 84, 86-90
    G3S(MP2), 79-81
    G3S(MP3), 79-81
    G3SX, 84, 86-88, 95
    G3SX(MP2), 87
    G3SX(MP3), 87-88

G3X, 68, 77, 81-86, 88-89, 94
G3X(MP2), 85-86
G3X(MP3), 85-86

Hartree-Fock (HF) approximation, 2, 3, 5-6, 9, 11, 14, 16, 18, 33, 35-37, 41-42, 50, 54-57, 59-61, 70, 77-78, 83, 85, 87, 102-107, 109, 112-114, 118-119, 122, 124-125, 132-137, 140, 147, 151-152, 155-156, 162-165, 178, 181-182, 190, 193, 200
Higher level correction (HLC), 68, 72, 76-79, 81, 83-86, 91-92, 163

Ionization potential, 46, 48-49, 51-52, 61, 69, 73-74, 80, 86-87, 131, 135-142, 147-155
IRCMax, 107-109, 125, 127

Koopmans's theorem, 134-135, 140, 142, 145

Møller-Plesset (MP) perturbation theory, 3, 162
MP2, 54, 71, 73, 76-77, 81-82, 84, 94, 100, 102-107, 109-112, 114-119, 124, 151-152, 162-165, 168, 173-182, 188-190, 193, 200-201, 203-207, 209-210, 213-215, 219, 221-228
MP3, 73, 102, 124, 162
MP4, 68, 71, 73, 78-79, 88, 103, 162-163, 178, 182, 190, 227-228
MP4(SDQ), 103, 119

ONIOM, 120, 122, 125

Polarizable continuum model (PCM), 120, 125
Pole strength, 134, 136, 140, 145, 156
Proton affinity, 46, 61, 69, 74, 80, 86-87

R12 theory, 15-20, 26, 109-112, 115-117
Radical stabilization energy (RSE), 161, 177-180
Reaction barrier height, 99, 107, 109, 124-125, 161, 181-189
Relativistic energy contribution, 23-26, 28, 34, 41-42, 54-55, 59, 61, 67, 166, 200, 202
mass-velocity term, 23-25, 34, 42, 56, 60
one-electron Darwin term, 23-25, 34, 42, 56, 60
spin-orbit term, 23-25, 34, 42, 50, 60, 72, 166
zero-order regular approximation (ZORA), 202

Semiempirical method, 235-243
AM1, 181-182, 190-191, 235-136, 238-242
MINDO/3, 235, 242

MNDO, 235-243
MNDOC, 235, 237, 242
MNDO/d, 235-236, 238-239, 241-242
MSINDO, 235-236, 242
OM1, 235-236, 239-241, 243
OM2, 235-236, 239-241, 243
PM3, 235-236, 238-242
SAM1, 235, 242
SAM1d, 242
SINDO1, 235, 242

Test set
G2/97, 48,50, 68-70, 72-82, 84-85, 89, 94-95, 145, 151-152, 154, 238-240
G2-1 subset, 42, 48-51, 56, 69, 73, 82, 90-91
G2-2 subset, 42, 48, 50-51, 56, 69, 82, 90-91
G3/99, 68-69, 73-74, 76, 80-82, 84-92, 94-95
G3-3 subset, 69, 75, 81-82, 89-91
W2-1, 41, 46-47, 50, 52, 54, 56

Weizmann-n (Wn) theory
W1, 31-41, 46-49, 51-52, 57-61, 118
W1′, 50-51, 163-165, 172, 174-176, 178-179, 181-182, 193
W1aug, 48
W1c, 47, 56, 61
W1ch, 56, 61
W1CAS, 57
W1h, 47, 49, 51-52, 56, 60-61
W2, 31-41, 46-52, 58-59, 61
W2CAS, 57
W2h, 47, 49, 51, 59-61

Zero-point vibrational (ZPV) energy, 9-10, 22-23, 26, 34, 43-46, 60-61, 70, 76, 82-83, 85, 87, 102, 107, 151, 152, 163-164, 166, 178, 182, 188-189, 208, 237-238

Understanding Chemical Reactivity

1. Z. Slanina: *Contemporary Theory of Chemical Isomerism.* 1986 ISBN 90-277-1707-9
2. G. Náray-Szabó, P.R. Surján, J.G. Angyán: *Applied Quantum Chemistry.* 1987 ISBN 90-277-1901-2
3. V.I. Minkin, L.P. Olekhnovich and Yu. A. Zhdanov: *Molecular Design of Tautomeric Compounds.* 1988 ISBN 90-277-2478-4
4. E.S. Kryachko and E.V. Ludeña: *Energy Density Functional Theory of Many-Electron Systems.* 1990 ISBN 0-7923-0641-4
5. P.G. Mezey (ed.): *New Developments in Molecular Chirality.* 1991 ISBN 0-7923-1021-7
6. F. Ruette (ed.): *Quantum Chemistry Approaches to Chemisorption and Heterogeneous Catalysis.* 1992 ISBN 0-7923-1543-X
7. J.D. Simon (ed.): *Ultrafast Dynamics of Chemical Systems.* 1994 ISBN 0-7923-2489-7
8. R. Tycko (ed.): *Nuclear Magnetic Resonance Probes of Molecular Dynamics.* 1994 ISBN 0-7923-2795-0
9. D. Bonchev and O. Mekenyan (eds.): *Graph Theoretical Approaches to Chemical Reactivity.* 1994 ISBN 0-7923-2837-X
10. R. Kapral and K. Showalter (eds.): *Chemical Waves and Patterns.* 1995 ISBN 0-7923-2899-X
11. P. Talkner and P. Hänggi (eds.): *New Trends in Kramers' Reaction Rate Theory.* 1995 ISBN 0-7923-2940-6
12. D. Ellis (ed.): *Density Functional Theory of Molecules, Clusters, and Solids.* 1995 ISBN 0-7923-3083-8
13. S.R. Langhoff (ed.): *Quantum Mechanical Electronic Structure Calculations with Chemical Accuracy.* 1995 ISBN 0-7923-3264-4
14. R. Carbó (ed.): *Molecular Similarity and Reactivity: From Quantum Chemical to Phenomenological Approaches.* 1995 ISBN 0-7923-3309-8
15. B.S. Freiser (ed.): *Organometallic Ion Chemistry.* 1996 ISBN 0-7923-3478-7
16. D. Heidrich (ed.): *The Reaction Path in Chemistry: Current Approaches and Perspectives.* 1995 ISBN 0-7923-3589-9
17. O. Tapia and J. Bertrán (eds.): *Solvent Effects and Chemical Reactivity.* 1996 ISBN 0-7923-3995-9
18. J.S. Shiner (ed.): *Entropy and Entropy Generation.* Fundamentals and Applications. 1996 ISBN 0-7923-4128-7
19. G. Náray-Szabó and A. Warshel (eds.): *Computational Approaches to Biochemical Reactivity.* 1997 ISBN 0-7923-4512-6
20. C. Sándorfy (ed.): *The Role of Rydberg States in Spectroscopy and Photochemistry.* Low and High Rydberg States. 1999 ISBN 0-7923-5533-4
21. P.G. Mezey and B.E. Robertson (eds.): *Electron, Spin and Momentum Densities and Chemical Reactivity.* 2000 ISBN 0-7923-6085-0

22. J. Cioslowski (ed.): *Quantum-Mechanical Prediction of Thermochemical Data.* 2001 ISBN 0-7923-7077-5

Kluwer Academic Publishers – Dordrecht / Boston / London

Zeitfracht Medien GmbH
Ferdinand-Jühlke-Straße 7
99095 Erfurt, Deutschland
produktsicherheit@kolibri360.de